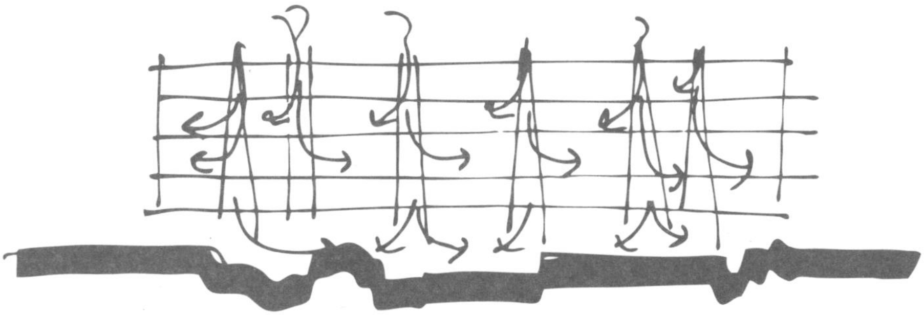

THE ARCHITECTURE & ENGINEERING OF DOWNDRAUGHT COOLING

A DESIGN SOURCEBOOK

EDITORS
BRIAN FORD, ROSA SCHIANO-PHAN, ELIZABETH FRANCIS

PHDC PRESS

Published by **PHDC Press** 2010

ISBN 978-0-9565790-0-3

Designed by **kirio comunicazione**, Bologna, Italy.

Printed and bound in the UK.

FOREWORD

It is a pleasure for me to introduce this Design Sourcebook on Passive and Hybrid Downdraught Cooling (PHDC) supported by the European Commission through the Sixth EU Framework Programme for Research and Technological Development.

The EU has set three commitments to be met by 2020, to reduce greenhouse gas emissions by at least 20 per cent, to ensure that 20 per cent of final energy consumption is met with renewable sources and to raise energy efficiency by 20 per cent. By 2050 the EU also intends to slash its greenhouse gas emissions by as much as 80 per cent. For this we need to develop and deploy technologies with no or low CO_2 *emissions. Passive Downdraught Cooling has a role to play. It consists of cooling living and working spaces with the mere use of air draughts. This requires appropriate designs but consumes zero energy, alleviating the use of air conditioning. This book will disseminate the findings of this project and multiply its impact.*

With these words, I invite the readers to take a benefit of this Design Sourcebook. It will be possible to learn a lot about the Passive Downdraught Cooling concept and its applicability in the built environment.

I wish everyone a stimulating read of this book!

Fabrizio Barbaso,
Deputy Director General for Energy,
European Commission

PASSIVE DOWNDRAUGHT COOLING: A DESIGN SOURCEBOOK
Project Partners and Authors

The material presented in this publication has been derived and developed from a global review of research and practice in the field of downdraught cooling, within the framework of the 'Passive and Hybrid Downdraught Cooling (PHDC)' Project, funded under the European Commission's 6th Framework for Research & Development. (Contract No.: TREN/07/FP6EN/S07.70683/0383418)

SIXTH FRAMEWORK PROGRAMME

Coordinator: **B.FORD** - University of Nottingham, UK
R.SCHIANO-PHAN - University of Nottingham, UK

Partners: **E. FRANCIS** - Mario Cucinella Architects, Bologna, Italy
S. ALVAREZ - AICIA, University of Seville, Spain
J. MOLINA - AICIA, University of Seville, Spain
J. SALMERON - AICIA, University of Seville, Spain
J. SPITERI-STAINES - Architecture Project, Malta
A. MICELI-FERRUGIA - Architecture Project, Malta
P. THOMAS - David Langdon Consultancy, UK
T. van STEENBERGHE - REHVA, Belgium
N. PATEL - Abhikram Architects, Gujarat, India
P. ZAVERI - Abhikram Architects, Gujarat, India
Q. XU - S.R.I.B.S., Shanghai, China

Sub-contractor: **E. ERELL** - Bengurion University of the Negev, Israel

Research Assistants: **M. ULLOA, J. WANG** - University of Nottingham, UK
C. GARAVELLI - Mario Cucinella Architects, Bologna, Italy
J. SANCHEZ - AICIA, University of Seville, Spain
G. DREYFUS - Architecture Project, Malta
A. GALLO - Architecture Project, Malta
A. AMATO - David Langdon Consultancy, London, UK
J. DING - S.R.I.B.S., Shanghai, China

Authors: Primer **(Brian Ford)**; Chapter 1 **(Brian Ford)**; Chapter 2 **(Brian Ford** with **Elizabeth Francis)**; Chapter 3 **(Brian Ford** & **Rosa Schiano-Phan)**; Chapter 4 **(Paul Thomas** & **Brian Ford)**; Chapter 5 **(Brian Ford)**; Chapter 6 **(Servando Alvarez, Jose Molina, Jose Salmeron** and **Brian Ford)**; Chapter 7 **(Brian Ford** with **Evyatar Errell)**.

Graphic Design: **www.kiriocomunicazione.it**

ACKNOWLEDGEMENTS

The authors would like to thank and acknowledge the assistance of:

Prof. Simos Yannas, AA Graduate School, London; Prof. Tadj Oreszczyn, UCL, London; Max Fordham, Max Fordham Partnership, London; Dr. Dejan Mumovic Faculty of the Built Environment, UCL; Prof. Derek Clements-Croome, University of Reading, UK; Dr. Malcolm Cook, Loughborough University; Prof. Alan Short, Short and Associates, London; Maria Widdowson, School of Slavonic and East European Studies, London; Adrian Leaman, Building Use Studies and Arup, London; Angela Clemo, UCL Estates, London.

Kevin Winschel, Building Services Technician at the General Services Administration, Phoenix, AZ; Keith Lew of FCU, Phoenix, AZ; Mahadev Raman, Arup New York; Chris Field, Director of the Department of Global Ecology, Stanford University; Linda Longoria, Department of Global Ecology. Stanford University; Ron Terry, Zion National Park; James Lutterman, Zion National Park; Paul Torvellini, Zion National Park.

Professor Li Baofeng, Wuhan University, China; Professor Joe Darkwa, CSET, Ningbo, China; Benson Lau and Huang Xuan, University of Nottingham, UK.

Dr Dutt, Director of the Torrent Research Centre, Ahmedabad; Meghna Meholta, Abhikram Architects, Ahmedbad; Karan Grover Architects, Baroda, India; Dr Prem Jain, Spectral Services; Ashish Rakheja, Surendra Shah, Sibasis Mohapatra; Ramesh Borad, Engineer and Systems Manager Torrent Research Centre; George Baird, University of Wellington; Lisa Thomas, University of Technology, Sydney; Prof. Arvind Krishan, School of Architecture & Planning, New Delhi; Prof. Vinod Gupta, Architect, New Delhi; Mr. Charlie Walker, British Council India and Sri Lanka.

Antonella Grossi, SAIE Bologna; Prof. Roberto Zecchin, Tiffs Ingegneria and University of Padua; Prof. Carlo Monti, University of Bologna; Mario Cucinella, MCA.

Eileen Muscat, Malta Stock Exchange; Steven De Bono, MTS Ltd, Malta; Helga Pizzuto, President of the Chamber of Engineers, Malta; Vincent Kassar, President of the Kamra Tal-Periti; Anthony Cardona, Malta Stock Exchange; John Ebejer, Building Industry Consultative Council, Malta.

Nina Maritz, Nina Maritz Architects, Namibia; Allison Kwok, Department of Architecture, University of Oregon; Walter Grondzik, College of Architecture and Planning, Ball State University; Bill Cunningham, Southwest Solar; Martin Yoklic, Environmental Research Laboratory, University of Arizona; The City of Melbourne.

Alexandros Tombazis, Tombazis Associates Architects; Evi Tzanakaki, Centre for Renewable Energy Sources, Greece; Ioannis Santzaklis, Centre for Renewable Energy Sources, Greece.

Energy Design Tools Team, UCLA;Renzo Piano Building Workshop, Italy; Wolfgang Motzafi-Haller, Blaustein Institute, Israel; Mr Srinivas, Principal Counsellor, Confederation of Indian Industries; Richard Meier Architects, USA.

Sergio Altomonte; Sailee Nalawade; Jeevan Mohan; Lyn Shaw, University of Nottingham.

Thanks also to Helen and Julia.

PREFACE
BACKGROUND

The construction industry is notoriously conservative and slow to change, and until recently, only the largest companies have engaged in research. However, concerns over global warming and the need for sustainable development, have led to substantial investment (particularly in Europe) into the research and development of technologies and design approaches which will reduce our dependency on fossil fuels. The application of these techniques to building projects has spread around the world, in parallel with the research programmes, and is slowly becoming part of the mainstream. Unfortunately, there is a perceived knowledge and skills gap among construction professionals, which is seriously restricting further take-up. There is widespread understanding of the 'need', but a lack of widespread 'know how'. The transfer of knowledge from research into practice is therefore vital. But this is difficult in an industry which habitually dislikes change and which has traditionally undervalued research.

In the last 20 years, developments in materials science, building energy technologies and design tools, have been derived from a significant increase in research investment into energy efficiency in buildings. Major advances have been made in both our understanding and the application of bioclimatic design principles in buildings, and similar advances have been made in new and renewable energy devices and their integration in buildings. Strategic thinking about these issues needs to be part of the early design process. Rigorous testing of options (using a wide range of analytic tools) is required in order to give clients the confidence to proceed with a radical solution. Not all projects can afford specialist testing and analysis, but simplified techniques exist, and this is where the 'know how' really counts.

It is important for us to look at the scale of the task that faces us, in terms of reducing carbon emissions and their impact on global warming. Atmospheric carbon today constitutes approximately 390ppm compared with the pre-industrial concentration of 280ppm. The projection of future carbon imbalance indicates that 'Business as Usual' is not an option. A strategy of avoiding fossil fuel use entirely would re-establish global carbon balance to pre-industrial levels, but even if this may be achievable technically, it is unlikely to be acceptable socially or economically. Current proposals suggest a target upper limit of 550 ppm by the middle of the century, although significant climate change is already apparent in many parts of the world, so the implications of this limit for the next century are rather alarming. For an energy strategy in buildings to be based on a combination of efficiency measures and the integration of renewables, both professional practice and education within the construction industry must change. The transfer of knowledge gained from recent research into practice must form part of this process. But in order for carbon emission targets to be achieved, gaps in knowledge and skills also need to be addressed among all who work in the industry. This Design Sourcebook on Passive and Hybrid Downdraught Cooling is intended as a contribution to this process.

1

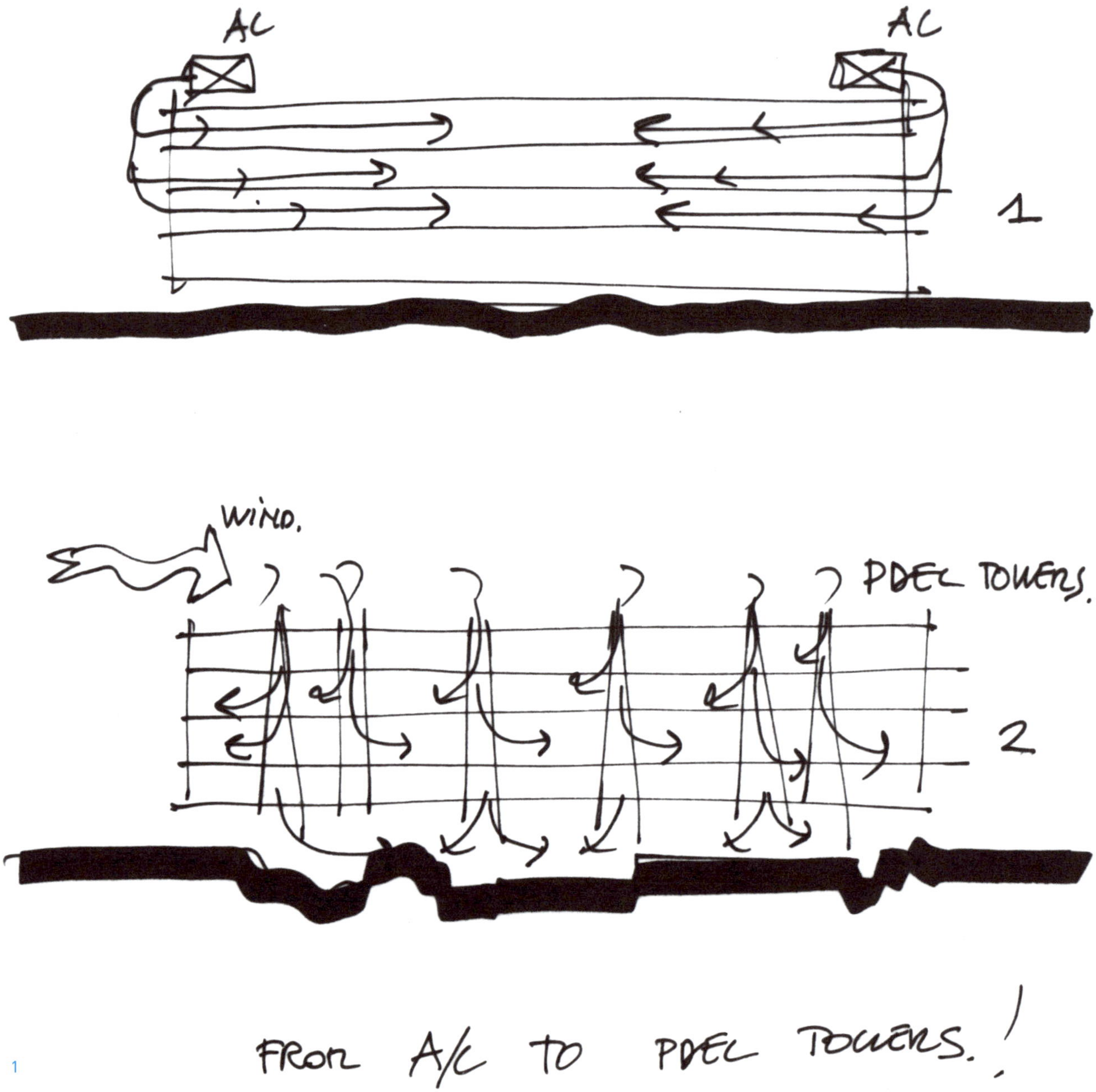

1. Sketch to show conceptual differences between air conditioning and PDEC as cooling strategies.

THE NEED FOR ALTERNATIVES TO CONVENTIONAL AIR-CONDITIONING

Buildings globally are responsible for about 40% of all energy consumption. CO2 emissions from air conditioning world-wide make a significant contribution to greenhouse gas emissions. During 2006, in the USA, air conditioning alone consumed approximately 11.5% of all electricity, but was responsible for over 40% of the peak load. The demand for air conditioning was a major reason for electricity supply problems on the west coast of the USA and in Italy in the exceptionally hot summer of 2003. While in Europe and the USA peak load demand for air conditioning threatens to disrupt supply, in India and China supply disruption is a regular occurrence, forcing many large consumers to invest in expensive back-up generating equipment. Alternatives to conventional air-conditioning are therefore urgently required to reduce peak load demand for electricity, and thus reduce the risk of supply disruption. Since the alternatives also exploit ambient heat sinks and are inherently 'low carbon', they will also reduce the need for investment in infrastructure.

The stratospheric ozone layer is the World's UV radiation shield. UV affects photosynthesis and kills phytoplankton (the basis of oceanic food chains). Refrigerants used in air-conditioning (both CFCs and HCFCs) are responsible for the 'holes' appearing in the ozone layer. Commonly used refrigerants (like R134a and R22) are between 1300 and 1700 times more potent per molecule than CO2 in contributing to global warming. CFCs are extremely stable and (in spite of the Montreal Protocol) are still being manufactured in China and India. HFCs continue to be used of course throughout the world although the search for less damaging alternatives is bearing fruit. Alternative refrigerants may be part of the answer, but there is now strong interest in more radical alternatives to air conditioning. PHDC is one of these alternatives.

This Design Sourcebook sets out the principles and practical opportunities in different parts of the world, and provides a global review of current practice, including 10 detailed case studies. Life-cycle cost benefits are also reviewed for Europe, China and India. In addition, new simplified performance prediction tools have been developed and are included on the CD with this Sourcebook.

CONTENTS

DESIGN PRIMER
ORIGINS AND OPPORTUNITIES

DOWNDRAUGHT COOLING: AN OPPORTUNITY

Downdraught cooling is an energy efficient, and cost effective alternative to conventional air-conditioning for new and existing buildings. As such, and in the context of rapidly rising demand for conventional air conditioning in buildings, it represents a significant opportunity for the reduction of carbon emissions from buildings.

It relies on the effect of gravity on a body of (relatively) cold air to create a downdraught, and thus circulate air from the source of cooling to the occupied zone within the building. The source of the cool air can be either 'passive' or 'active'.

Passive downdraught cooling. A passive downdraught can be achieved through the evaporation of water within an air-stream. This is referred to as Passive Downdraught Evaporative cooling or 'PDEC', which is only appropriate in hot dry conditions. In southern Europe, passive downdraught evaporative cooling can meet 25 – 85% of the cooling load of non-domestic building (equivalent to 15 - 60kWh/m2 pa). In residential buildings the cooling load can be reduced to well below 15 kWh/m2 pa.

Active downdraught cooling. In warm and humid conditions active downdraught cooling can be achieved by using chilled water cooling coils or panels. While relying on mechanical cooling, this avoids the need for fans, which can represent an energy saving of 25-35% of the electrical load in a non-domestic building. This approach also avoids the need for either bulky fan-coil units or air-handling units and related ductwork.

Hybrid downdraught cooling. A hybrid downdraught cooling system combines both 'passive' and 'active' downdraught cooling techniques. Such a system can function in both hot and dry conditions (using PDEC) and warm and humid conditions (using chilled water cooling coils).

IN THIS PRIMER WE ARE CONCERNED WITH PASSIVE AND HYBRID DOWNDRAUGHT COOLING OR 'PHDC'

In this Primer we are concerned with Passive and Hybrid Downdraught Cooling or 'PHDC'.

Conventional cooling systems use fans to drive the airflow to the occupied zones of a building. Downdraught cooling systems differ from conventional cooling systems in that the primary mechanism for distribution of cool air is either by buoyancy or wind assisted natural ventilation. This means that both energy costs and capital costs can be significantly reduced.

Typically the capital cost of a PDEC / Hybrid cooling system is comparable in cost with comfort cooling, whereas full air conditioning is significantly more expensive. Expressed as a percentage of new build costs, PDEC is in the order of 4-12% compared with 19 - 23% for full AC installations. The percentages are approximately doubled as proportions of the cost of refurbishment work. Detailed lifecycle cost analyses for Europe, China and India indicate that Passive & Hybrid Downdraught Cooling has significant life cycle cost benefits, typically providing a return on investment within a few years (2-5 years in India; 5-15 years in Europe and China).

DOWNDRAUGHT COOLING AVOIDS THE NEED FOR FANS, WITH SIGNIFICANT LIFE CYCLE COST BENEFITS

ORIGINS

The tradition of 'Cooling without Air-Conditioning', which incorporates a range of design responses to climate, has its origins in ancient Egypt. It subsequently spread eastwards as part of the Islamic tradition through the Middle East and Iran to north India (with the Moghul empire), and westwards across North Africa to southern Spain with the Moors. (fig. 2)

This tradition, which has been largely overlooked, is characterized by dramatic achievements. Travelling through the Iranian desert on his journey to China in the 13th Century, Marco Polo commented on his being offered fruit flavoured water-ices to relieve the summer heat. The creation and storage of ice in the desert was made possible at the time through the construction of huge natural refrigerators. Under the clear night sky of the desert, surface temperatures drop low enough to freeze a thin layer of water. This was then cut up and stored in a massive domed ice house. In this way, the extremes of the environment were turned to the advantage of the people living there. This reflects an attitude of respect and symbiosis between people and their environment, working with it rather than against it, to enhance the quality of their lives.

IN THIS TRADITION, WIND-CATCHERS GUIDE OUTSIDE AIR OVER WATER-FILLED POROUS POTS, INDUCING EVAPORATION

This understanding is elaborated further in a long tradition in the Middle East of using various techniques to encourage air movement and evaporative cooling both within and between buildings. Water jars mounted in specially designed window openings in Muscat, Oman, cool the air passing over them into the room. In a similar manner, woven 'Khus' matting suspended over window and door openings are still commonly used in parts of northern India.

Passive cooling and ventilation of buildings in Iran, (incorporating wind catchers, porous water pots and salsabil) have been widely applied and very effective for several cen-

2 

3 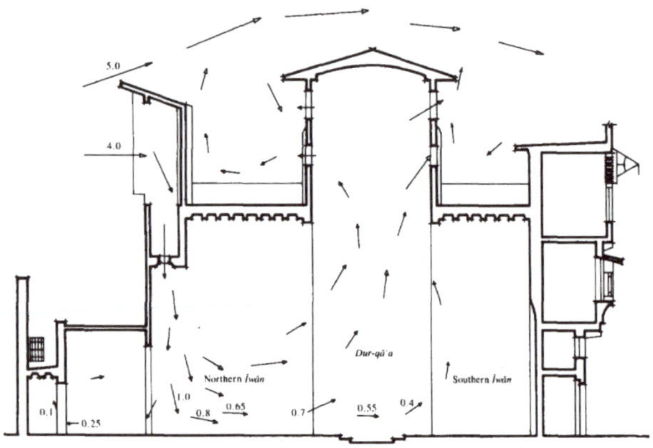

2. Iranian wind catcher.

3. Airflow in an Egyptian malqaf, studied by scholars from the AA School of Architecture in London.

IN NORTH INDIA THE MUGHAL PALACE AND GARDENS EXPLOITED EVAPORATIVE COOLING TO DELIGHT THE EYE AND OTHER SENSES AS WELL AS PROVIDING THERMAL RELIEF

turies. In this tradition, wind-catchers guide outside air over water-filled porous pots, inducing evaporation and bringing about a significant drop in temperature before the air enters the interior. Hassan Fathy was also aware of this tradition in Egypt, adapting, developing and re-applying these techniques to cool and ventilate schools and housing projects. In a field study which involved students from the Architectural Association, Fathy reports on measurements of temperature and air velocity within a house with 'malqafs' (windcatchers), illustrating the pattern of airflow and the benefits of enhancing air movement within the building. (fig. 3)

In north India the Mughal palaces and gardens exploited evaporative cooling to delight the eye and other senses as well as providing thermal relief. Thin water chutes ('salsabil'), and other evaporative cooling techniques were features of Mughal architecture from the 13th to the 17th centuries. (figs. 4 a - b - c)

4a-b

4c

4. (a) Water chute with niches for oil lamps, (b) sloping 'salsabil', and (c) salsabil and lotus pool in Red Fort Red Fort, Delhi, India.

The intense dry heat and dust of the summer in north India calls for the creation of an internal 'refuge' or haven from the extremes of the external world. The diurnal swing in temperature is 'dampened' by the mass of stone and earth, and the air is further cooled by the evaporation of water in the ventilation air flow path. This is exemplified perfectly in the beautifully atmospheric summer room in the 'Rai Praveen Mahal' in Orchha, Madhya Pradesh. Built towards the end of the 16th Century, the main rooms in the house look north over an orchard and garden. On the north side, underneath a raised terrace, is a room partially sunk into the ground and flanked by two large open water tanks. As the visitor descends into this elegant room, the relief from the intense summer heat and the peace and tranquillity of the space, leaves a deep impression. Although the pools are empty now, the original intention of the architect is still clear: small

THE FORM AND ARCHITECTURAL EXPRESSION OF THESE BUILDINGS IS DERIVED FROM THE NEED TO RESPOND TO THE CULTURAL AND CLIMATIC CONTEXT

openings on each side of the space with deep reveals to prevent solar gain, encourage the movement of evaporatively cooled air across the space. Even without water, a 10°C temperature difference was found between inside and outside, and the addition of evaporative cooling to the air stream would have made this room comfortable in the most extreme heat of summer. (figs. 5 - 6 - 7 - 8)

The form and architectural expression of these buildings derived from the need to respond to the cultural and climatic context while creating a habitable and comfortable environment in what were often extreme conditions.

Nearly 50 years ago, Reyner Banham was beginning to piece together a very novel form of architectural history, which was to become a classic of its time. 'The Architecture of the Well Tempered Environment', chronicled the impact which the development of building services engineering had on the design of contemporary architecture, and the nature of the architect's role in the design team.

One of the key developments which has had a profound influence on architecture, was the invention by Willis Carrier (just over a hundred years ago in 1906), of mechanical air-conditioning. Banham was the first to reveal that one of the bi-products of the air-conditioned interior was that the form and fabric of the building envelope had ceased to function as the primary moderator of the external climate on the internal environment. In the process, many architects lost their knowledge of, and interest in, the environmental control functions of the building itself. Indeed, in many contemporary buildings the external envelope only serves to exaggerate dependency on mechanical conditioning to achieve a satisfactory environment. This of course has been a gradual process, and many of the great architects of the twentieth century proved to be exceptions to this rule. However, by the end of the 20th Century, the control of the internal environment had come to be regarded as almost entirely the province of the engineer.

But times are changing. Many (architects and engineers alike) believe that it is no longer acceptable for engineers to have to make good by brute force what others have failed to achieve by design. There is renewed interest in the rich, varied and subtle vocabulary of an architecture which successfully moderates the impact of the external environment on the internal environment to achieve thermal as well as visual delight, without total dependence on mechanical intervention. Alternatives are being sought for conventional air-conditioning, which is now recognized as a significant contributing factor to global warming and climate change. This search has led to a re-discovery of the principles of environmental control through the manipulation of the building form and fabric: essentially 'passive' design principles which enable cooling without air-conditioning.

Architects are now turning to ancient techniques, such as downdraught cooling for application in contemporary buildings. Both Fathy and Bahadori sought to make improvements to the traditional windcatchers of Egypt and Iran. Often, the air flow rate is enhanced by catching and redirecting the prevailing wind using methods similar to traditional wind towers, or increasing surface area to enhance evaporation.

The need to reduce energy demand for cooling and the search for alternatives to conventional air-conditioning, led to experimental buildings in different parts of the world. Experiments in Phoenix at the University of Arizona represent the

ONE OF THE BI-PRODUCTS OF THE AIR-CONDITIONED INTERIOR WAS THAT THE FORM AND FABRIC OF THE BUILDING ENVELOPE HAD CEASED TO FUNCTION AS THE PRIMARY MODERATOR OF THE EXTERNAL CLIMATE ON THE INTERNAL ENVIRONMENT

5

7

8

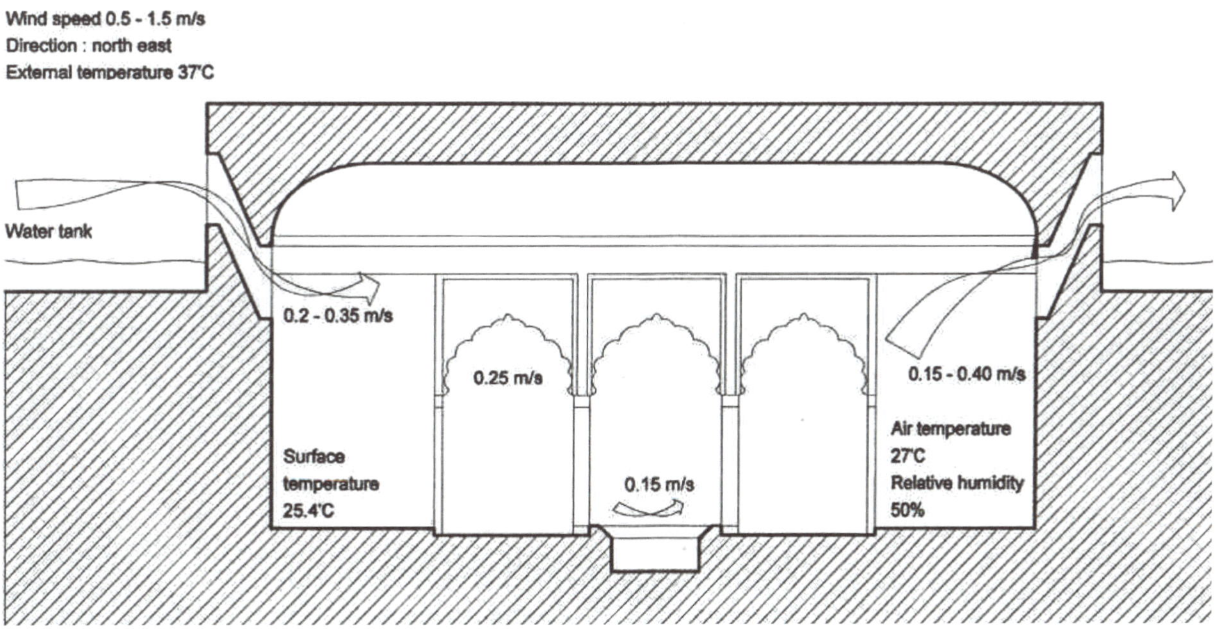

6

5. General view of Rai Praveen Mahal, Orchaa, India.

6. Section of summer room, Rai Praveen Mahal, showing air flow pattern and relationship with evaporative cooling ponds.

7. Rai Praveen Mahal.
Image of the summer room in 1994.

8. Painting of Rai Praveen.

THIS SEARCH HAS LED TO A RE-DISCOVERY OF THE PRINCIPLES OF ENVIRONMENTAL CONTROL THROUGH THE MANIPULATION OF THE BUILDING FORM AND FABRIC

9a

9b

9c

9. (a-b-c) Misting Towers, Avenue of Europe, Seville Expo '92.

first contemporary application of these techniques to cool the interior of a building, and demonstrate that passive downdraught cooling can achieve very high air change rates, as well as achieving acceptable internal temperatures.

Many aspects of the historic tradition outlined above were also reviewed by the designers of the Expo site in Seville for the 1992 World Fair. This included the 30 metre high 'cool towers' of the Avenue of Europe, which employed high pressure water misting nozzles (micronisers) to induce downdraught cooling. (figs. 9 a - b - c)

The application of misting nozzles within a tower serving a prototype office building was tested as part of the PDEC research project at the laboratories of Conphoebus in Catania, Italy. This experimental building, and the detailed design proposals which followed, demonstrated that when the ambient wet bulb temperature is low enough, PDEC can achieve satisfactory conditions within office buildings in southern Europe. The project also demonstrated that the performance of PDEC could be predicted using a combination of CFD and dynamic thermal modelling. (figs. 10 - 11)

In parallel with this research and experimentation, a number of buildings have now applied the technique of passive downdraught evaporative cooling in its various forms. While some of these buildings are not without their faults, they have demonstrated that passive and hybrid downdraught cooling is technically and economically viable now, and that it has enormous potential to displace the need for conventional air conditioning.

PASSIVE AND HYBRID DOWNDRAUGHT COOLING IS TECHNICALLY AND ECONOMICALLY VIABLE NOW

10

11
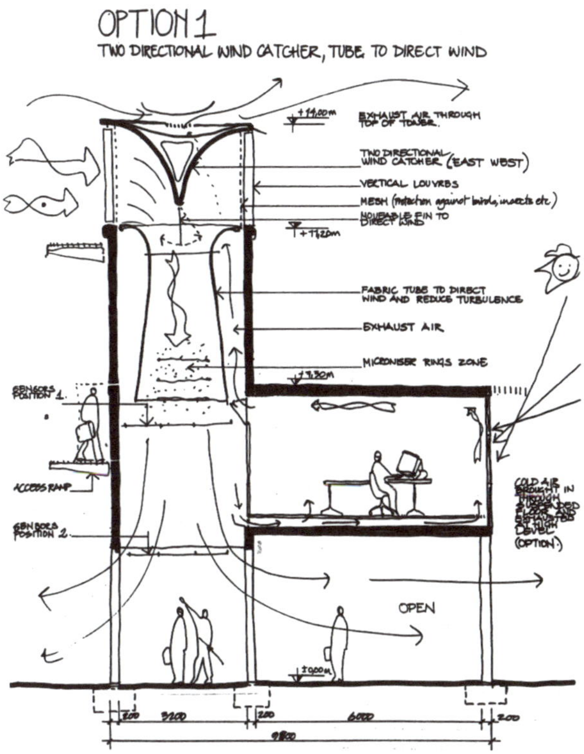

10. Experimental PDEC Building by MCA at Conphoebus, Catania, Italy.

11. Section and air-flow strategy in experimental PDEC Building by MCA at Conphoebus, Catania, Italy.

APPLICABILITY

Hybrid downdraught cooling can be applied in warm and hot climates, either humid or dry and is therefore viable in a very wide range of locations.

Where the external wet bulb temperature goes above 24°C then PDEC is not appropriate. In order to understand whether a particular location is suitable, monthly climate data can be plotted on a psychrometric chart to reveal the general viability of evaporative cooling.

A hybrid system is required in humid conditions, when the downdraught can be achieved using cooling coils. This is mechanical cooling, but avoids the need for fans and associated ductwork, to achieve a significant energy saving plus savings in capital and maintenance costs.

A wide range of building types are amenable to the integration of downdraught cooling systems, including commercial and industrial, residential, educational, health and large volume buildings such as transport interchanges.

Downdraught cooling has been applied to both new and existing buildings in Europe, the USA, India, Australia, the Middle East and Africa. Assessments have been made of the wider applicability of downdraught cooling to existing commercial and residential buildings in Europe, India and China.

A study of the application of PDEC to existing commercial buildings in southern Europe revealed that PDEC is applicable to potentially 70% of this stock, (saving 3.7 - 5.7 m tonnes of carbon per annum). Analysis of applicability to the housing stock in Spain, Portugal, Italy and Greece has shown that the residential market for PDEC is even larger, with potential annual savings of 19 - 38m tonnes of carbon.

HYBRID DOWNDRAUGHT COOLING CAN BE APPLIED IN WARM AND HOT CLIMATES, EITHER HUMID OR DRY

SYSTEMS

A downdraught of evaporatively cooled air may be generated by four different methods, which serve as a means of classifying the different system options.

Cool Towers

The irrigation of a cellulose matrix, or wetted pads, in the path of the air stream.

Shower Towers

Droplets of water sprayed into the air stream.

Porous media

Wetted porous surfaces located within the air stream, for example porous ceramics.

Misting Towers

A fifth option is a hybrid system, which relies on a passive downdraught (no fans), but an active source of cooling (cooling coils).

A WIDE RANGE OF BUILDING TYPES ARE AMENABLE TO THE INTEGRATION OF DOWNDRAUGHT COOLING SYSTEMS

Cool Towers

In this option wetted cellulose pads are incorporated at the top of the air inlet tower. They promote a large drop in temperature, and high air change rates (typically 10-20 air-changes per hour). This combination can drive the ventilation air through the building. A number of small non-domestic and residential buildings in the hot dry states of south-western USA have been designed incorporating cool towers with wetted cellulose pads. (figs. 12 - 13)

Shower Towers

In this case showerheads at the top of the tower induce the evaporation of large droplets of water within the air-stream. This has been used in a number of buildings in Australia, the Middle East and the USA. Much of the water does not evaporate and has to be collected in a pool at the base of the tower. The advantage is that water quality can be relatively poor, but the disadvantage is that the collection pools occupy a proportion of the floor plan. (figs. 14 - 15 - 16)

Porous media

In many parts of the world where water quality is poor, or where the use of misting nozzles is regarded as too 'high tech', the use of porous ceramic panels to promote evaporative cooling may be appropriate. The re-interpretation of this traditional method was the subject of an EC funded research project, and building projects are beginning to apply this technique. (fig. 17)

12. Zion National Park Visitor Centre.

13. Zion National Park Visitor Centre. Cooling strategy.

12

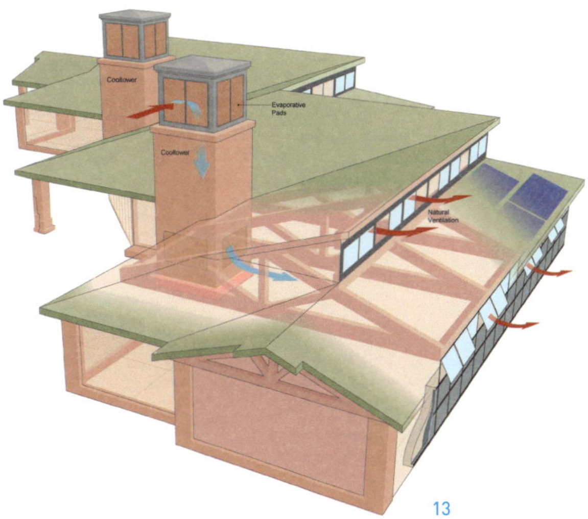

13

14

15

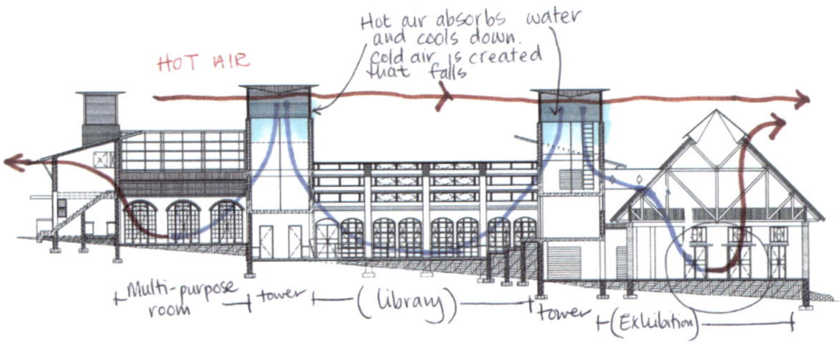

16

17

18

14. The Habitat Research & Development Centre (HRDC), Namibia. General view.

15. Shower tower. The Habitat Research & Development Centre (HRDC), Namibia.

16. Section of HRDC and cooling strategy.

17. Sectional model showing 'Evapcool' strategies.

18. Torrent Research Centre. Ahmedabad, India.

Misting Towers

The amount of cooling achieved through evaporation increases as droplet size decreases. Misting nozzles have been used for many years in botanical gardens, and in the food processing and chemical industries. The first large scale outdoor application of the technology in southern Europe was in the 'Avenue of Europe' at the Seville Expo in 1992. Now the use of these nozzles in external cafes in southern Europe is commonplace.

The first large scale application within a building was in the Torrent Research Centre at Ahmedabad in India where misting towers have successfully provided cooling to pharmaceutical research laboratories and offices since 1998. (fig. 18) Recent developments in misting nozzle technology now allow evaporation at low pressure, making this the most efficient and cost effective system of downdraught evaporative cooling.

Hybrid Systems

Where the external wet bulb temperature goes above 24°C some form of back-up to a PDEC system may be required. Conventional chilled water cooling coils located high in the supply air shafts, can perform this role, and have been applied in the Malta Stock Exchange. (figs. 19 a - b - c)

19. Malta Stock Exchange uses a hybrid strategy, using (a) PDEC, (b) cooling coils and (c) night time convective cooling.

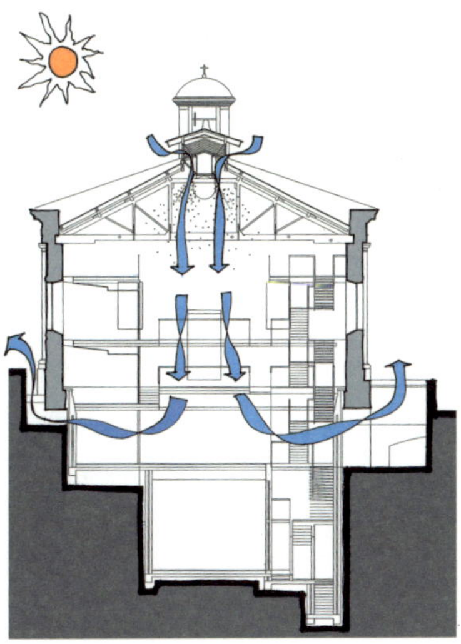

19a

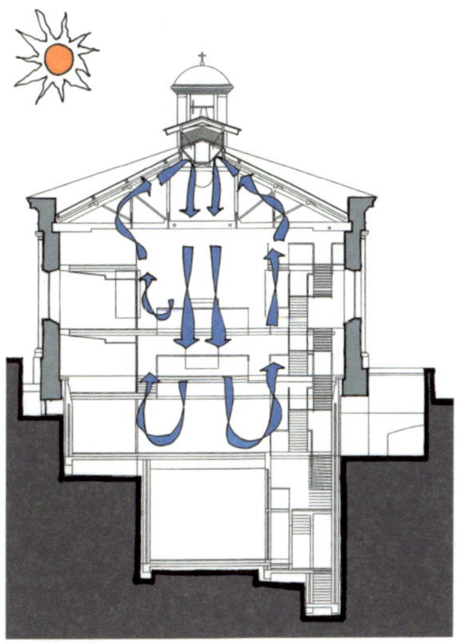

19b

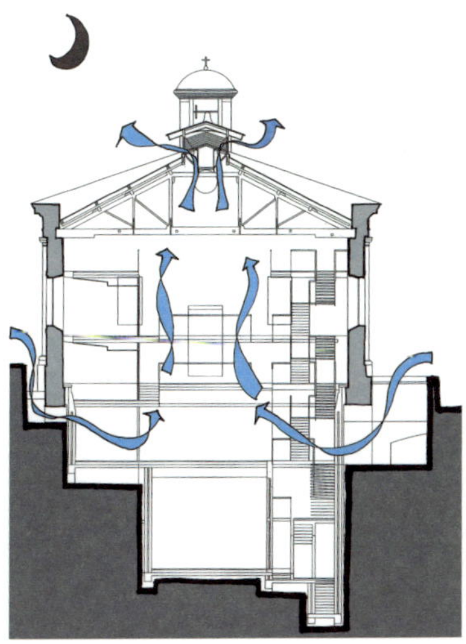

19c

PERFORMANCE

Where misting nozzles within a tower are used to provide downdraught evaporative cooling, then:

- Average electrical energy savings of between 25 – 60 kWh/m2 (a 50 – 80% reduction of the cooling and associated fan energy loads in non-domestic buildings in southern Europe).
- In residential buildings the residual cooling load is brought well below 15kWh/m2 pa. In many circumstances, supplementary cooling in residential buildings may be avoided altogether.
- Air temperatures 8 – 12°C below ambient dry bulb, plus high air change rates, providing thermal comfort for occupants for much of the working day.

The micro-climate and the expectations of occupants influences whether additional cooling is required. In the Torrent Research Laboratories in Ahmedabad, occupant perceptions are very favourable, indicating that, PDEC can provide satisfactory conditions without supplementary cooling.

Performance simulation tools (predicting temperature reduction, air flow rates and energy savings) are available for the different evaporative cooling systems. Simple monograms are also available to obtain broad-brush initial indications of the likely performance of Cool Towers.

COMPONENTS

All downdraught cooling applications require:

- Either an inlet shaft, tower, light-well or atrium to deliver the cool air directly to spaces within the building.
- Baffled openings, shafts or ventilated facades to ensure that air is then exhausted successfully (irrespective of wind direction or pressure).

In non-domestic buildings, these elements replace the air handling units, ductwork and false ceilings of conventional systems. In residential buildings the cool air may be introduced via stairwells or light shafts, and exhausted via baffled openings. The geometry of the building is therefore a major determinant of the downdraught cooling system.

20. Rai Praveen Mahal.
Mural of the Dancing Queen.

20

ECONOMIC CONTEXT

While the nature of the construction and property market, and the costs of construction vary significantly, the economic implications of Climate Change and 'Peak Oil' will be similar around the world. Rising electricity prices will make PHDC increasingly attractive.

Reliable cost data exists for a few benchmark buildings. The installation in the Malta Stock Exchange cost 115Euro/m2 treated floor area for PDEC, or 177 Euro/m2 including chillers and cooling coils (2003). In the Seville building the PDEC installation cost 77 Euro/m2, or 96 Euro/m2 including chillers and cooling coils (i.e. a 'Hybrid' system).

Analysis of construction costs within Europe suggests that typically a PDEC / Hybrid cooling system is comparable in cost with comfort cooling, whereas full air conditioning is significantly more expensive. Expressed as a percentage of new build costs in Europe, PDEC is in the order of 4-12% compared with 19 - 23% for full AC installations. The percentages are approximately double as proportions of refurbishment work. Analysis of life cycle costs for both new build and existing buildings (i.e. refurbishment) suggests that PHDC will pay for itself within 5 – 15 years.

It has been established that, within southern Europe, PHDC represents a technically viable alternative to conventional air-conditioning, and the market is likely to grow. Projections of the market for PHDC in non-domestic buildings in southern Europe concluded that: By 2025, some 11% of both annual refurbishments and annual new build production would incorporate PDEC interventions.

ANALYSIS OF CONSTRUCTION COSTS WITHIN EUROPE SUGGESTS THAT TYPICALLY A PDEC / HYBRID COOLING SYSTEM IS COMPARABLE IN COST WITH COMFORT COOLING, WHEREAS FULL AIR CONDITIONING IS SIGNIFICANTLY MORE EXPENSIVE

CULTURAL CONTEXT

There is a long history of evaporative cooling in different parts of the world. However, its acceptability now may depend on expectations with regard to environmental control within buildings, and also attitudes to traditional solutions. They may be regarded as 'old fashioned' or not 'progressive', and therefore undesirable. This was the reaction to Hassan Fathy's early work in Egypt. However, the ranges of buildings around the world applying this approach are extremely varied in their expression, and demonstrate that PHDC presents an opportunity for innovation.

Expectations regarding the nature of the internal environment of buildings vary significantly in the different regions of the world. Differences in expectations may also explain why building occupants in the USA appear to be more critical of their internal environment, than their counterparts in India, for example. Our own surveys of user satisfaction tend to support this. In certain locations these issues may represent a barrier to the take up of downdraught cooling.

THE RANGE OF BUILDINGS AROUND THE WORLD APPLYING THIS APPROACH ARE EXTREMELY VARIED IN THEIR EXPRESSION

GEOGRAPHICAL CONTEXT

There are predictions that global warming will increase the occurrence of extreme weather conditions, including more severe droughts and floods. We can expect higher summer-time temperatures wherever we are in the world, and with the growing urban population, the urban heat island effect will have an additional impact. The demand for cooling is therefore likely to rise significantly. In Europe, demand for air-conditioning is predicted to double by 2020, and the market is expanding at an even faster rate in India and China. Low-carbon alternatives to conventional air-conditioning are urgently required. Downdraught cooling is one of these alternatives.

CHAPTER 1
PRINCIPLES AND STRATEGIES

1.1 SITE & MICRO-CLIMATE ANALYSIS
The impact of climate

> THE CHARACTERISTICS OF ALL SEASONS ARE EQUALLY IMPORTANT IN DETERMINING AN OVERALL PASSIVE DESIGN STRATEGY FOR THE BUILDING

The local ambient conditions and climatic characteristics (temperature, relative humidity, solar radiation, wind speed and direction, etc) are clearly of primary concern when considering the risk of overheating for a any location, and in determining the viability of a specific passive cooling strategy. Climate data can be obtained for a particular location by reference to either local meteorological station (airports have traditionally been a location for recording weather data), or from international data bases like 'Meteonorm'[1].

From the point of view of cooling we are primarily interested in the summer period. However, the characteristics of the other seasons are equally important in determining an overall passive design strategy for the building. A graph of the monthly temperature and relative humidity variation, or a plot of the same data on a psychrometric chart can help to reveal the opportunity (and limitations) of an evaporative cooling strategy (see section 2.3). The applicability of PHDC in Spain is discussed as an illustration.

Detailed climate data was sourced from Spanish Weather for Energy Calculations (SWEC) and the range in temperature was analysed using Climate Consultant 4[2]. The figures below show a general description of Madrid climate. (fig. 21)

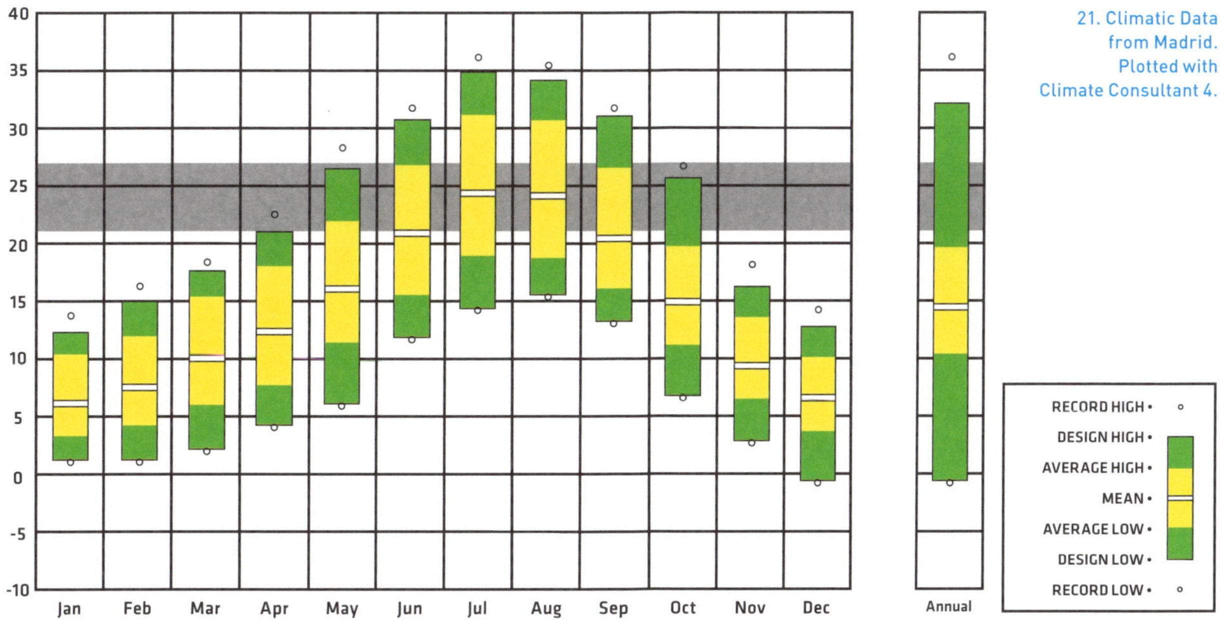

21. Climatic Data from Madrid. Plotted with Climate Consultant 4.

21

Psychrometric chart analysis

Plotting data for Madrid on a psychrometric chart reveals that during the summer period (June – September) thermal comfort (assuming a dry bulb air temperature between 22 – 25°C) should theoretically be achievable by evaporative cooling alone. This is because although the dry bulb temperature is high (above 35°C on occasion), the relative humidity is consistently low (generally below 50% RH during the day).

The weather data for Madrid is also characterised by a large diurnal range, with night time temperatures below 20°C during the summer months. This suggests that night time convective cooling can also be exploited to precool the building interior for the start of each day (see section 1.2).

During the mid-seasons of Spring and Autumn, mean monthly temperatures of 12 – 16°C suggest that comfort can be maintained within the house by natural ventilation alone. Occupant control of window openings should enable regulation of natural ventilation to remove any heat gains during the day, while set to 'trickle' ventilation at night to provide minimum fresh air. (fig. 22)

22. Psychrometric chart for Madrid, Spain.

22

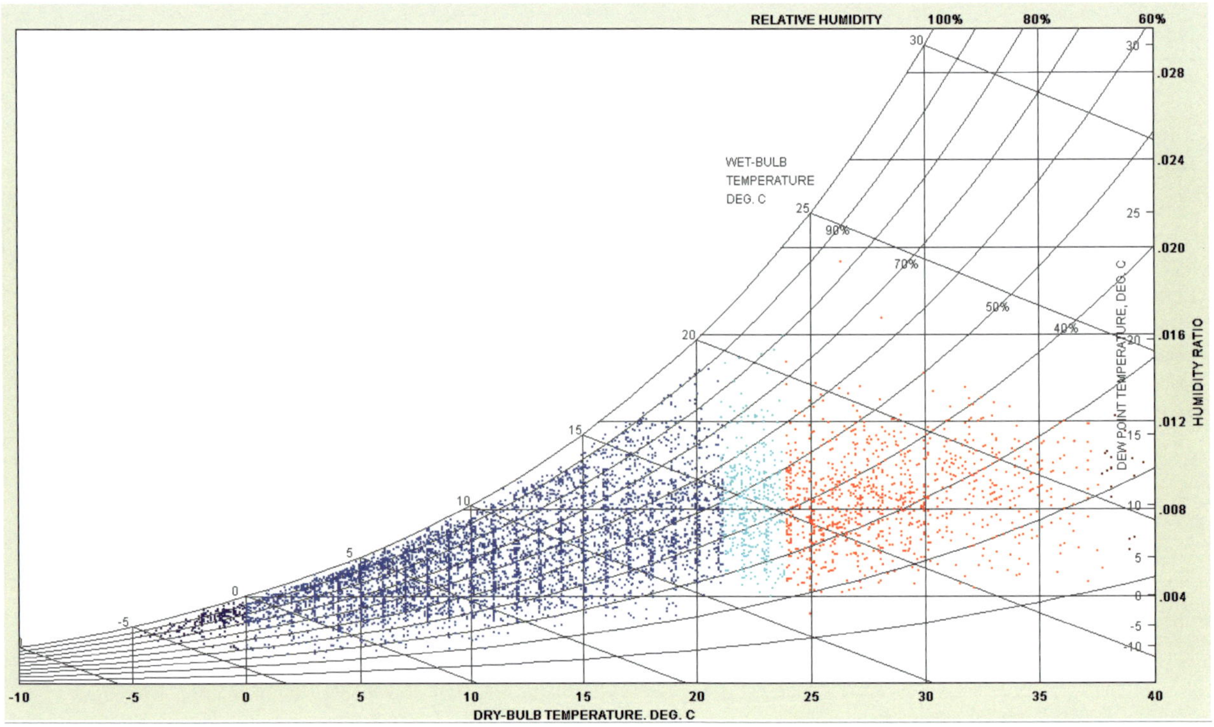

The physical context

The meteorological data needs to be interpreted in the context of the particular physical characteristics of the site. The wind regime and solar exposure will of course be influenced by the proximity of other buildings, trees or topographical features, and may then either enhance opportunities of the site or inhibit them. Equally, the proximity of roads (which may be a source of noise and pollution) or green spaces (quiet clean areas) may have an influence on strategic thinking with regard to the location of ventilation openings into and out of the building.

Distinctions may also be made between urban, sub-urban and rural locations. Urban microclimatology has become a field of significant research over the last fifteen years, and much has been written on the subject[3]. Urban morphology also has a significant influence on the micro-climatic context of a single building. This can be illustrated by contrasting the densely packed high rise buildings of Hong Kong, where individual buildings overshadow and block the wind from their neighbours[4], and the low rise sprawl of Los Angeles, where the wide boulevards, dark roof surfaces and omnipresent air-conditioning contribute to a significant heat island in the city.

DISTINCTIONS MAY BE MADE BETWEEN URBAN, SUB-URBAN AND RURAL LOCATIONS

Street Pattern

The 'urban grain' – the pattern and hierarchy of streets, their orientation, and the relative width, height and density of surrounding buildings – can potentially have a significant impact on the local micro- climate. The orientation of streets will define when sunlight falls on different elevations and whether wind is obstructed or channelled through a particular urban area. The micro-climatic characteristics of the surrounding area can help to frame an environmental design response. (fig. 23)

Urban Heat Islands

Urban heat islands are characterised by significantly elevated temperatures relative to the surrounding urban fringe or rural areas. Areas of generally higher average temperatures are associated with denser urban areas with dark surface colours and lacking green areas. In Los Angeles, where a study by MIT in the 1990s revealed the urban heat island effects, new building codes require roofs to be white in colour[5]. Parks, gardens and green areas can contribute to lower average temperatures, and studies have shown that there can be significant variation in temperature within a small area[6]. These local urban heat island variations can have a big impact on cooling load.

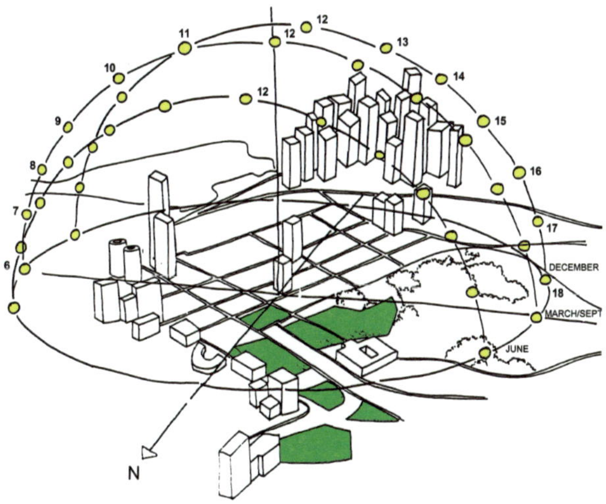

23. The relationship between urban morphology, street pattern and sunpath for Singapore.

23

Effects of Vegetation

Evapo-transpiration (the absorption of CO2 and the release of water vapour by trees and other vegetation) can reduce temperatures locally[7]. This effect is most pronounced where the vegetation is contained between buildings or within courtyards. This effect has long been recognised and exploited in hot dry regions[8]. A recent study of the Casa de Pilatos in Seville, Spain, indicates that in summer, temperature differences of up to 10°C occur between 'green' courtyards and dry paved courtyards.

Traffic Noise and Air Borne Pollution

Traffic noise and air borne pollution may be a challenge, but not necessarily an impediment to natural ventilation and passive cooling. Their occurrence around a particular site should be mapped, as it may impact on the environmental design strategy. Analysis of noise and pollution for a particular site will reveal where sound baffling may be required and also where the cleanest fresh air can be obtained to supply the building. Passive design solutions can be as viable in city centre locations as they are in suburban and rural areas.

Wind Effects

Wind can be exploited to promote ventilation and cooling of building interiors. Prevailing wind speeds and direction need to be determined over the year to identify opportunities and constraints, depending on the nature of the building type and other characteristics of the site. Suburban and rural sites will generally be exposed to higher average wind speeds than city centres in the same regional location. Wind 'catchers' have been employed historically to enhance airflow through buildings in exceptionally hot climates, and this can be an effective strategy today, depending on building type and location[9]. Design strategies must account for variations in wind pressure around the building, depending on wind speed and direction and the surrounding topography. Examples of spatial variation of wind pressure coefficients for rectangular buildings are given in *Numerical data for air infiltration and natural ventilation calculations*[10]. (fig. 24)

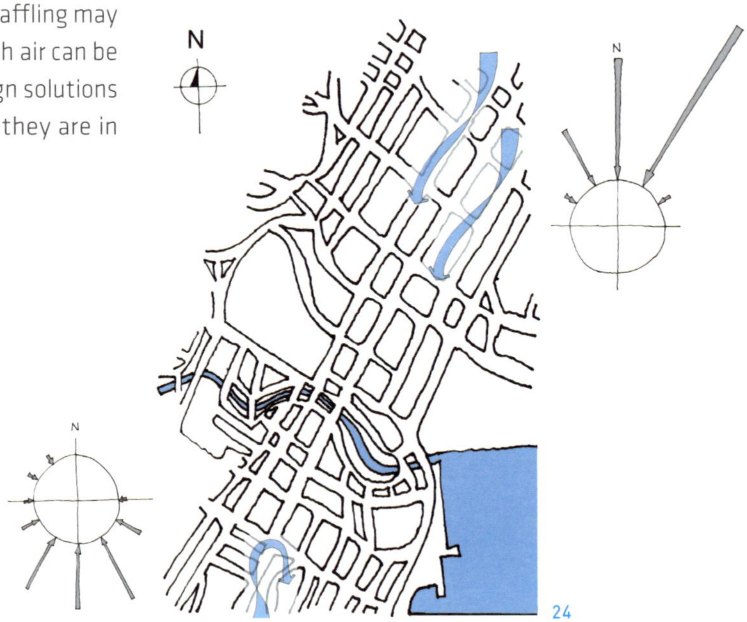

24

24. The relationship between street pattern and prevailing winds for central Singapore.

1.2 PASSIVE COOLING DESIGN STRATEGIES

'Passive' cooling refers to the exploitation of an environmental heat sink to achieve cooling[11] (as opposed to 'active' cooling which relies on mechanical intervention). Environmental heat sinks include night time external air, the night sky, the ground and the evaporation of water. Heat from within a building can be removed through the movement of air or water between the building and the heat sink.

There is a conceptual difference between conventional cooling and cooling based on environmental heat sinks. In conventional systems, the air is usually cooled by exchanging energy with cool water (produced in the chiller). The heat goes to a refrigerant and is finally transferred to the environment thanks to a compressor. The opportunities of developing design strategies to exploit environmental heat sinks are discussed below.

HEAT FROM WITHIN A BUILDING CAN BE REMOVED THROUGH THE MOVEMENT OF AIR OR WATER BETWEEN THE BUILDING AND A HEAT SINK

Night Ventilation

Throughout Europe, a relatively large diurnal swing in air temperature is experienced in summer, with night time temperatures dropping well below the 'neutral' temperature. This cool air can be drawn into the building to flush out any residual heat from the day and to pre-cool the internal fabric for the following day. (figs. 25 - 26) The coupling of the air flow path with internal components with high thermal diffusivity promotes convective heat transfer from internal surfaces to the air (convective heat transfer depends not only on the amount of energy stored (capacity) but also upon the conductivity of the material. (Diffusivity = the ratio between the thermal conductivity and the volumetric heat capacity) This air is then exhausted via automatically controlled vent openings to regulate both the air flow rate and the temperature inside the building, promoting pre-cooling prior to the arrival of the buildings occupants the next day. This technique is now widely used in both domestic and non-domestic buildings, wherever there is sufficient diurnal temperature variation[12-13].

THROUGHOUT EUROPE, A RELATIVELY LARGE DIURNAL SWING IN AIR TEMPERATURE IS EXPERIENCED IN SUMMER

25. SFC Brewery Process Building, Malta.
South elevation.

25

26. SFC Brewery Process Building, Malta.
Night ventilation strategy.

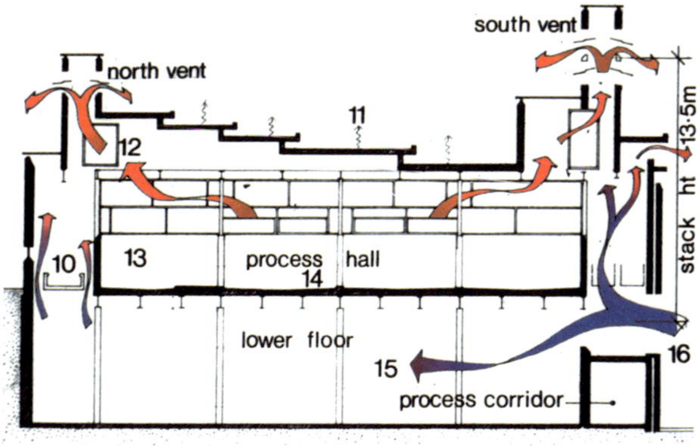

26

Night Sky Radiation

Hot dry regions of the world often will have clear skies at night, which can provide a potential heat sink by radiation from the relatively warm surface of the roof of a building, to the cold upper atmosphere. A simple and commonly applied technique is to circulate water from the interior to the roof (e.g. the Lowara office building in Italy by Renzo Piano Building Workshop, (figs. 27 a - b) and the Global Ecology Centre at Stamford University). A range of techniques for exploiting night sky radiation, including irrigated roofs and roof-ponds, are described in 'Roof Cooling Techniques' by Yannas, Erell and Molina[14].

27a

27. (a-b) Lowara Office.
Irrigated roof
(©Renzo Piano Building Workshop).

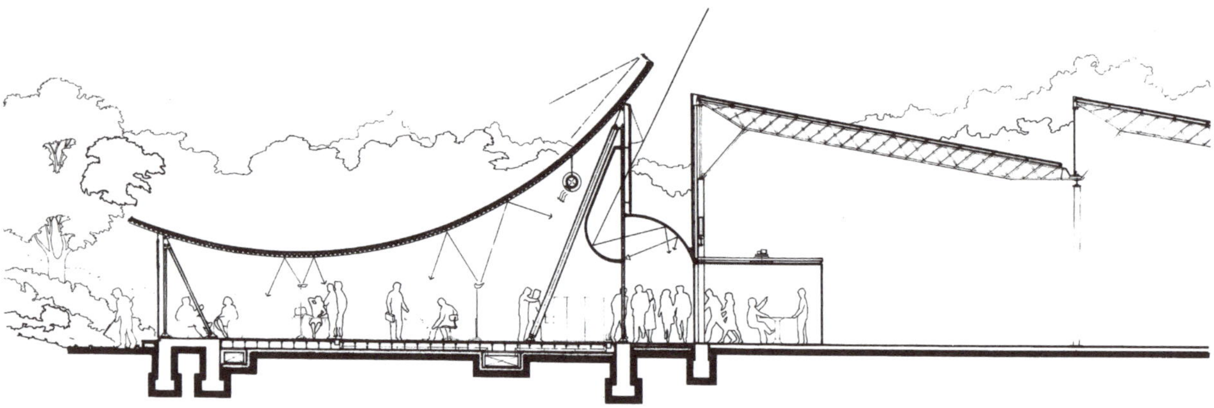

27b

Ground Cooling

The temperature of the Earth ten metres below ground level is generally very stable, and has been found to be equal to the annual mean air temperature for the location (anywhere in the world), varying perhaps by ±2°C according to the season[15-16]. The earth is therefore a huge source of low grade heat, which can be used for either heating or cooling. (figs. 28 - 29 - 30) For example, the annual average air temperature for central England is about 12°C. By introducing air to the building through a network of buried pipes, or via an underground labyrinth, it can be pre-cooled in summer and pre-heated in winter. This source of heating and cooling can also be exploited by circulating water though pipes in the ground and then exchange heat through a heat pump to either increase or decrease the temperature according to the season, before delivering heat or 'coolth' via intermediate floor slabs within the building.

THE EARTH IS A HUGE SOURCE OF LOW GRADE HEAT, WHICH CAN BE USED FOR EITHER HEATING OR COOLING

28

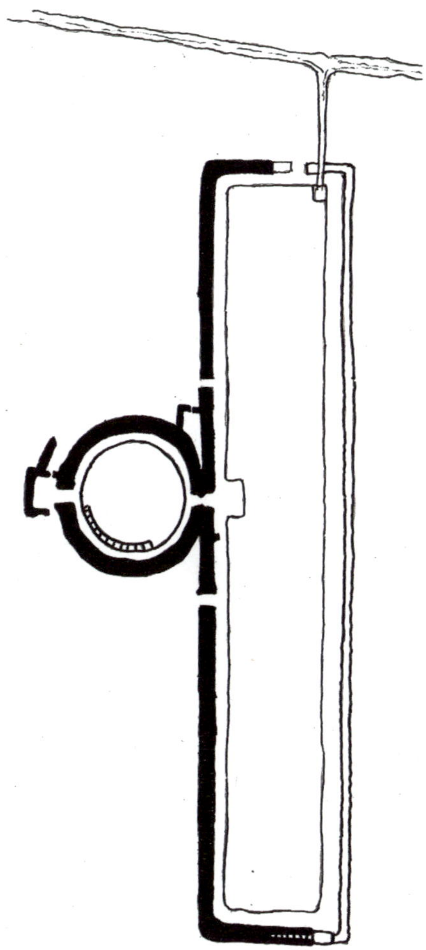

30

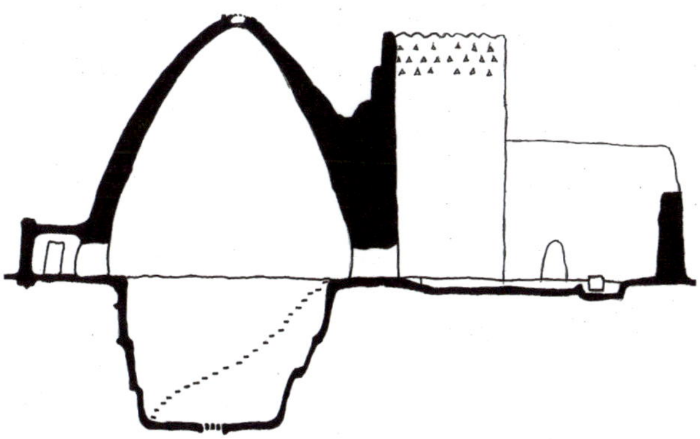

29

28. Ice House near Yazd, Iran, general view.

29. Ice House near Yazd, Iran, section.

30. Ice House near Yazd, Iran, plan.

Evaporation

EVAPORATIVE COOLING OCCURS AS A RESULT OF THE LATENT HEAT OF VAPORIZATION

The cooling potential of evaporating water has been exploited throughout southern Europe, the Middle East and northern India for many centuries. As a general rule, a reduction in supply air temperature of 70- 80% of the dry bulb-wet bulb depression can be achieved. This of course varies from only a few degrees when the air is relatively humid, to as much as 10 to 12°C or more, when the air is relatively dry. The upper limit for the application of direct evaporative cooling has been suggested by Givoni to be a wet bulb temperature of 24°C[17]. However, adaptive comfort opportunities may influence the acceptable upper limit significantly. As with convective cooling, control of the rate of evaporation and the air flow through ventilation openings is vital to optimize performance and to avoid over-humidification.

Evaporative cooling occurs as a result of the latent heat of vaporization – the amount of energy that is required to convert unit mass of water from its liquid to its gaseous/vapour phase (~2,260kJ·kg-1). This is appreciably greater than the specific heat required to raise the temperature of unit mass of water (4.18kJ·kg-1). In evaporative cooling, the energy is supplied by ambient air whose heat content and capacity to hold vapour in moist air are indicated by its dry bulb temperature and relative humidity. The transfer of heat energy from ambient air to water supports the phase change process and so reduces the temperature of the ambient air[18]. For a given pressure, the wet bulb temperature depends on the air temperature and the humidity content of the air. For example, for typical summer conditions of Seville, (38°C, 30% relative humidity), the wet bulb temperature is 23.5°C.

There are a number of different strategies which can be adopted to exploit either direct or indirect evaporative cooling. We are concerned in this Sourcebook with direct passive evaporative cooling, and related 'hybrid' systems which incorporate buoyancy-driven mechanical cooling. (fig. 31)

31

31. Evaporative cooling misting nozzles in action.

1.3 CONTEMPORARY APPLICATIONS

Downdraught cooling relies on the effect of gravity on a body of (relatively) cold air, to create a downdraught, and thus circulate air from the source of cooling to the occupied zone within the building. The source of the cool air may be either 'passive' (e.g. through the evaporation of water) or 'active' (e.g. from chilled water cooling coils or panels). The downdraught may be augmented (or inhibited) by wind effects, but is often designed to rely on buoyancy alone.

As we have seen in the Design Primer, Passive Downdraught Evaporative Cooling (PDEC) involves the creation of a cold downdraught of air by evaporating water within an air-stream. The downdraught of evaporatively cooled air may be generated by either:

- the irrigation of a cellulose matrix in the path of the air stream (e.g. wetted pads in a 'Cool Tower').
- large droplets of water sprayed into the air stream (e.g. by a 'Shower Tower')
- a mist of water sprayed into the air stream (e.g. by misting nozzles in a 'Misting Tower')
- wetted porous surfaces located within the air stream (e.g. porous ceramic in an 'Air Shaft').

Cool Towers

"the irrigation of a cellulose matrix in the path of the air stream"

In an experimental building in Tucson, Arizona in 1986[19] a downdraught tower incorporating wetted cellulose pads demonstrated the effectiveness of direct evaporative cooling in driving a substantial airflow through the building (figs. 32 - 33). In the extreme dry climate of Arizona, the potential for effective cooling by direct evaporation is very significant. Givoni, in analysing the test data, confirmed the effectiveness of this strategy[20]. From a sample of the published results it is interesting to note that the exit air temperature from the cooling tower was measured at 24°C when the outdoor dry bulb temperature (DBT) reached 41°C and the wet bulb temperature (WBT) was 22°C. The large drop in temperature through the cooling tower, coupled with high air change rates (in this case believed to be induced by buoyancy alone), indicates not only the significant cooling potential of this technique, but also that it drives the ventilation air through the building. Since this experimental work, a number of buildings in south-west USA have been designed incorporating cool towers with wetted cellulose pads. (fig. 34)

Cunningham & Thompson's experiments demonstrated that buoyancy forces alone can achieve very high air change rates (30 air changes per hour were recorded in their experimental building). This is very significant because it implies that fans, which are nowadays used to drive air around a building, can be avoided. Fan power can represent 25% - 35% of the electrical energy required by a conventionally air conditioned building. The innovation in relation to contemporary buildings is therefore not the use of evaporative cooling per se, but rather the rediscovery that it can drive the air flow through the building. Typically, in a new air conditioned office building in southern Europe this might represent a saving of 70-80kWh/m2 per year (35-45kg CO_2/m2 per year).

While this represents the potential savings, cellulose pads within cool towers (fig. 35) provide a significant resistance to airflow, and therefore large surface areas of pads are required in relation to the volume flow rate (see sizing procedures in Chapter 5). Water consumption can also be high as there are substantial losses to ambient, and water is supplied to the pads for most of the day. (fig. 36)

A NUMBER OF BUILDINGS IN SOUTH-WEST USA HAVE BEEN DESIGNED INCORPORATING COOL TOWERS WITH WETTED CELLULOSE PADS

32. General view of Experimental Building, University of Arizona, Phoenix, USA by Cunningham & Thompson, 1989.

32

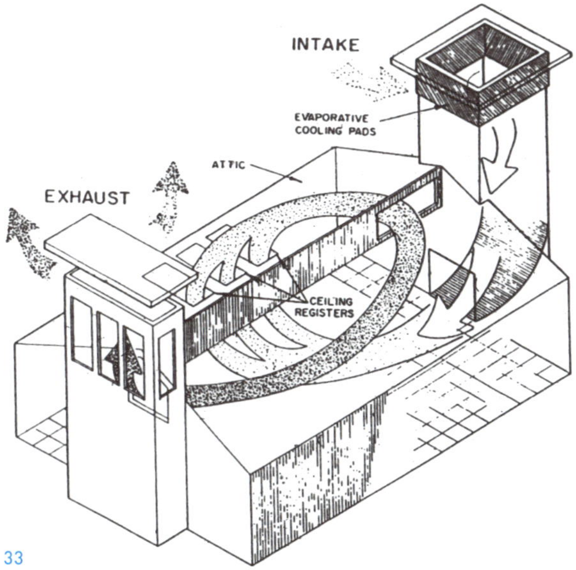

33

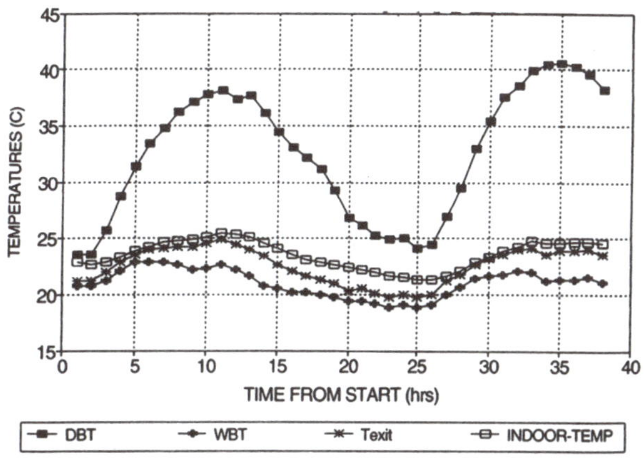

34

35

36

33. 3D diagram of Experimental Building, University of Arizona, Phoenix USA, by Cunningham & Thompson, 1989.

34. Temperature data from Experimental Building University of Arizona, Phoenix USA by Cunningham & Thompson, 1989.

35. Cellulose mats and spray nozzles mounted in test rig, Seville, Spain 1991.

36. Cool Towers at Zion National Park Visitors' Centre, Utah, USA .

Showers Towers

"large droplets of water sprayed into the air stream"

To overcome some of the disadvantages of using cellulose pads, the direct evaporation of droplets of water within the air-stream has been adopted as a strategy in a number of buildings. Hassan Fathy had experimented with running water over charcoal within an Egyptian 'Malqaf'[21], collecting the water in a pool at the base of the tower. The use of simple shower heads at the top of the tower within a small courtyard was also tested experimentally, first by Givoni and Al-hemiddi in southern California[22-23], and then by Etzion et al[24] in southern Israel (Negev). In the latter case, the shower tower serves an enclosed (glazed) courtyard, adjacent to an office building and library. The use of large droplet shower heads means that much of the water does not evaporate, and has to be collected in a pool at the base of the tower. The influence of wind catcher design on air volume flow rate was compared with fan induced air movement. The performance of the Israeli project has been recorded over a number of years and reported[25]. This project was also the subject of a post-occupancy evaluation study in 2008, and is described in the Case Studies (Appendix A). (fig. 37)

37

37. Shower tower in Blaustein Institute,Israel.

Misting Towers

"a mist of water sprayed into the air stream"

Many aspects of the tradition outlined above were reviewed by the designers of the Expo site in Seville for the 1992 World Fair[26]. This included the 30 metre high 'cool towers' of the Avenue of Europe, which employed high pressure water misting nozzles to induce downdraught cooling.

The first large scale application within a building was in the Torrent Research Centre at Ahmedabad in India, where misting towers have successfully provided cooling to pharmaceutical research laboratories and offices since 1998. Since then, a number of other buildings in India and the USA have applied this approach. Experience in the design of these buildings suggests that the successful application of misting towers requires a Void to Floor ratio of 0.05 (i.e. where 5% of the building footprint would be void). These voids normally take the form of vertical shafts, light-wells or atria, incorporating misting nozzles located at least two metres above the top floor served. Automatic control of inlet and outlet vents is required at each level. These vents must also be sized to balance the airflow between floors, and baffles or exhaust stacks may be required to prevent counter-flow.

MISTING TOWERS HAVE SUCCESSFULLY PROVIDED COOLING TO PHARMACEUTICAL RESEARCH LABORATORIES AND OFFICES SINCE 1998

At the Stanford Ecology Centre, where evaporative cooling is used to cool the reception space, the misting nozzles were installed at the top of the downdraught tower.
(figs. 38 - 39 - 40 - 41)

38

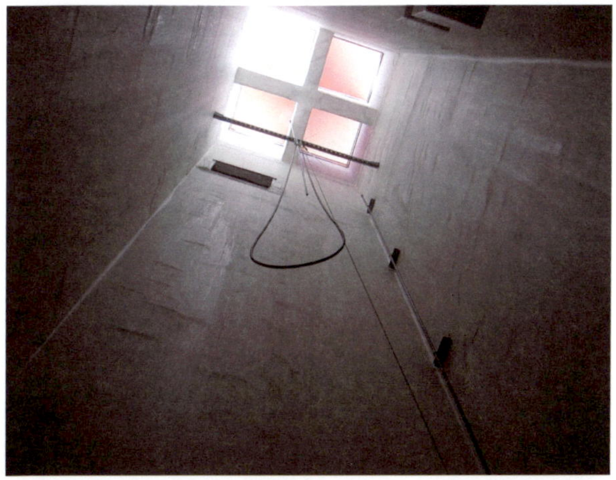
39

40

41

38. Stanford Ecology Centre. General View.

39. Stanford Ecology Centre. Downdraught Tower.

40. Stanford Ecology Centre. Reception area receives cool air from downdraught tower.

41. High pressure misting nozzle.

Porous Ceramics

"wetted porous surfaces located within the air stream"

The historic use of porous ceramic jars to promote evaporative cooling can be reinterpreted today to provide cooling to individual (cellular) spaces. The production of a cool downdraught by the wetting of porous ceramic bricks within a tower has been shown to be a viable design approach in Greece[27]. The use of porous ceramic panels within a wall element was also tested as part of the EC funded EVAPCOOL research project (The final report of the EVAPCOOL Project is included in its entirety on the accompanying CD), which enabled the performance characteristics to be defined. The applicability of this technique to housing in southern Spain has also been studied, but has so far not been applied in practice. A local authority office building in Tehran, Iran was designed to incorporate 'Evapcool' wall panels, but was modified at construction stage to incorporate an irrigated porous wall, the performance of which is currently being reviewed. The 'wicking' effect of porous ceramic components was exploited in the Spanish Pavilion at the Zaragoza Expo 2008[28]. While this technique has yet to be adopted in a significant demonstration project this is certainly a fertile area for further development.

The design opportunities, airflow strategies and prediction of performance of the different downdraught evaporative cooling techniques outlined above are discussed in more detail in section 1.4.

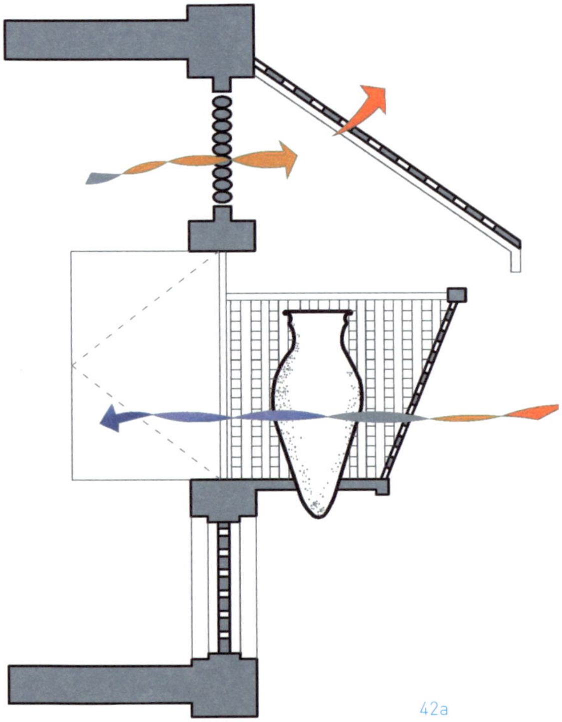

42a

42b

42. (a) Window mounted porous jar for evaporative cooling Muscat, Oman.

42. (b) "Botijo", a porous ceramic jar used in Southern Spain.

Buoyancy Driven Mechanical Cooling

While passive downdraught evaporative cooling may be applicable in much of southern Europe, it is rarely applicable for 100% of the time (simply because at times the wet bulb temperature is too high). Therefore, some form of back-up cooling may be required, depending on location, building type, and internal heat gains. In other climatic regions (warm humid or cool temperate), evaporative cooling may never be appropriate due to continually high humidity. In seeking alternatives to conventional a/c systems, a back-up or stand alone system that mimics the downdraught operation of PDEC has considerable advantages. Chilled water cooling coils located high in an atrium or light-well, have been used successfully in a number of buildings to perform this role. Chilled water cooling coils located at the top of

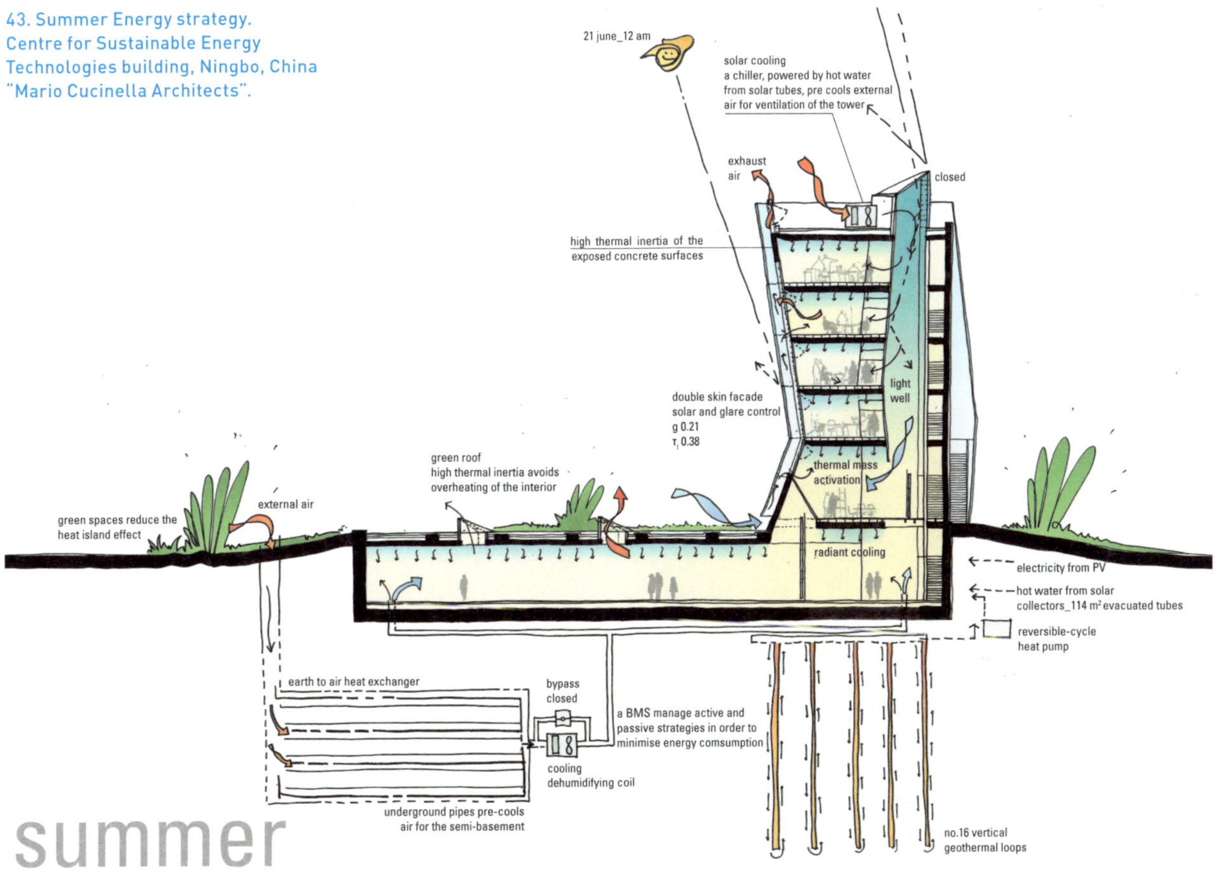

43. Summer Energy strategy. Centre for Sustainable Energy Technologies building, Ningbo, China "Mario Cucinella Architects".

IN SEEKING ALTERNATIVES TO CONVENTIONAL A/C SYSTEMS, A BACK-UP OR STAND ALONE SYSTEM THAT MIMICS THE DOWNDRAUGH OPERATION OF PDEC HAS CONSIDERABLE ADVANTAGES

a column of air, can induce a downdraught of cool air into a space, displacing warmer air which rises to the height of the coils before being cooled again, thus setting up a natural buoyancy driven circulation of air. (figs. 44 a - b) This can also be induced within a room (single floor height) or within a large volume (many floors in height). By avoiding the need for a full mechanical cooling system (including air handling units, ductwork, suspended ceilings etc), significant capital and energy savings can be achieved. While requiring chilling equipment, the avoidance of fans, ductwork etc also reduces maintenance and replacement costs. Room based systems such as 'Gravivent'® and 'G-Therm'® are available in the market.

Such an approach has been applied to a wide range of building types in different parts of the world, but rarely in conjunction with passive downdraught evaporative cooling. Both techniques exploit buoyancy to drive the airflow, avoiding the need for fans. They can be combined successfully to meet cooling needs where ambient conditions vary from hot & dry to hot and humid. Buoyancy driven mechanical (cooling using conventional cooling coils) has recently been applied in the Malta Stock Exchange and the School of Slavonic and Eastern European Studies, for UCL London (described in detail in Appendix A). Buoyancy driven cooling has also been adopted in the CSET building at the University of Nottingham campus in Ningbo, China, but in this instance the cool air is introduced into the top of the shaft via a fan driven de-humidifier. Buoyancy driven mechanical cooling has also recently been applied in the Malta Stock Exchange.

44. (a-b) School of Slavonic and East European Studies, London. Short Associates Architects.

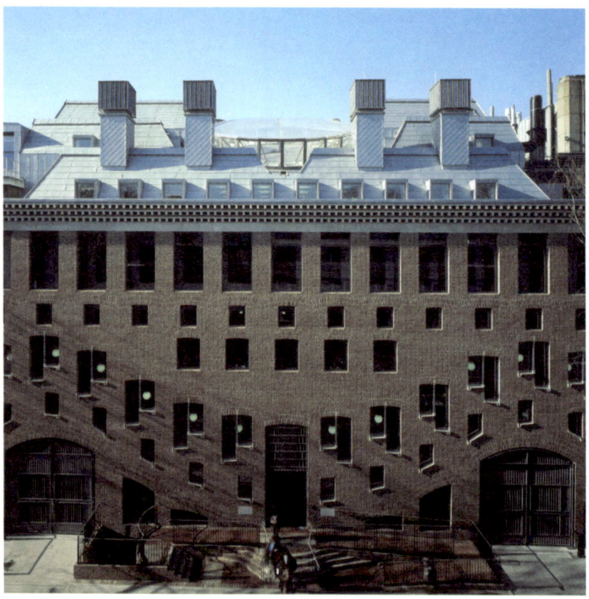
44a

44b

1.4 AIRFLOW STRATEGIES GEOMETRIC OPTIONS

The airflow strategy can have a significant impact on building geometry, and therefore needs to be considered early in the design process. There are a number of different geometric options which define the relationship between the source of PDEC supply air and the served spaces. The different options result in different airflow paths, and may influence cooling effectiveness and comfort. The four variants of Passive Downdraught Evaporative Cooling can be classified according to the mode of delivery/generation of cool air: Misting towers, Shower Towers, Cool Towers and Wall Shafts. For each delivery mode there are also different geometric options which will influence the pattern of airflow.

Supply air from a misting tower can either be delivered at the centre of the plan or from the perimeter. We have identified four generic cases:

- Case 1 - Central Atrium or Concourse (Open)
- Case 2 - Central Shaft (Enclosed)
- Case 3 - Perimeter Tower (Attached)
- Case 4 - Perimeter Tower (Detached).

All four cases assume buoyancy driven airflow, and that the impact of the wind is inhibited by baffles. Opportunities for exploiting the wind to increase the air volume flow rate are discussed separately below. It should also be noted that in all cases (apart from Case 4) the airflow pattern can be reversed at night to promote night time convective cooling.

CASE 1
Central Atrium or Concourse (Open) Misting Tower

An open central atrium or concourse (fig. 45) can be considered as a single cell, where the supply zone and the delivery zone are the same, and surrounding spaces are not served by the downdraught cooling system. Case study buildings of this type include the Phoenix Courthouse and the Malta Stock Exchange (see Chapter 3). For initial calculations, the temperature in this single zone can be assumed to be uniform. However, in practice, temperatures can vary significantly within an atrium served by downdraught cooling. This may not be a problem, since atria are often used as transition spaces, and can be allowed to have a wider range of temperature and humidity than surrounding work spaces. However, where the atrium serves as work or recreation space, then potential variations need to be minimized.

In a competition proposal for the Johannesburg Constitutional Court, a PDEC system was proposed to cool the large central public space during the day in summer, while reversal of the airflow pattern takes place at night, promoting convection cooling. (figs. 47 - 48)
The open geometry of an atrium or concourse also supports the incorporation of chilled water cooling coils (mounted at high level) as a back-up cooling source for times when the relative humidity is too high for direct evaporative cooling to operate effectively. The upper threshold for the internal relative humidity is normally set at 65%. (This is discussed more in the sections on thermal comfort and controls). This case assumes buoyancy driven air flow, and the impact of the wind on the supply tower is inhibited by baffles, to prevent wind blowing evaporatively cooled air straight out of the tower. The design of wind baffles is also discussed in more detail below (Chapter 7).

45. Atrium of the Malta Stock Exchange.

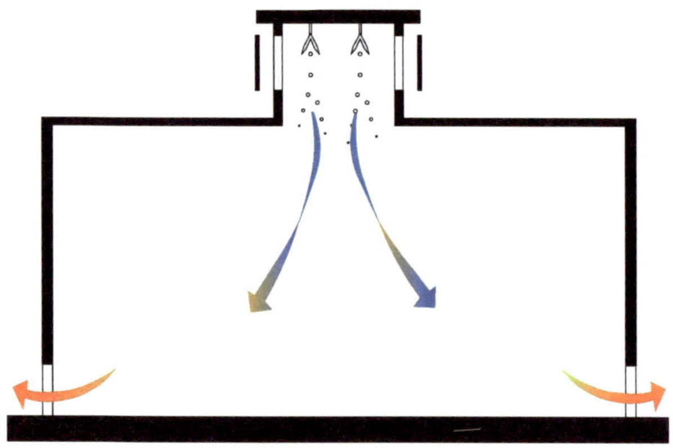

46. Case 1. Central atrium or concourse diagrammatic section.

47. Competition for Constitutions Court, Johannesburg. PDEC Strategy.

48. Competition for Constitutions Court, Johannesburg. Night ventilation.

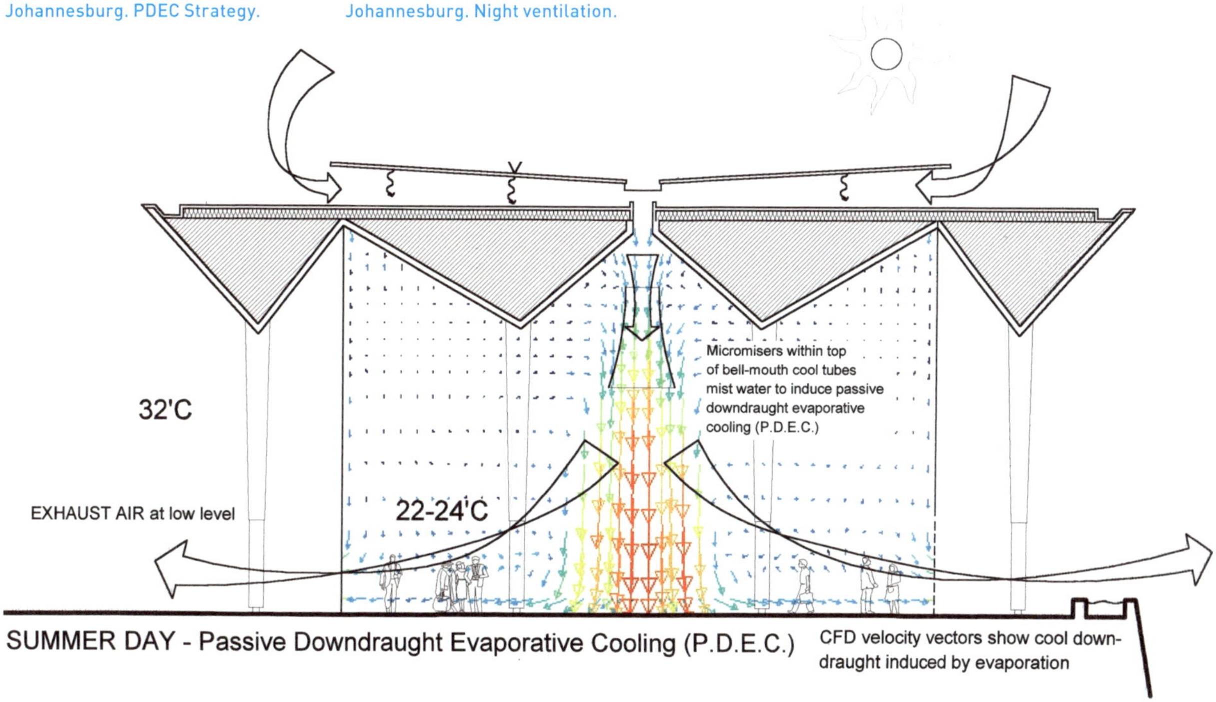

47

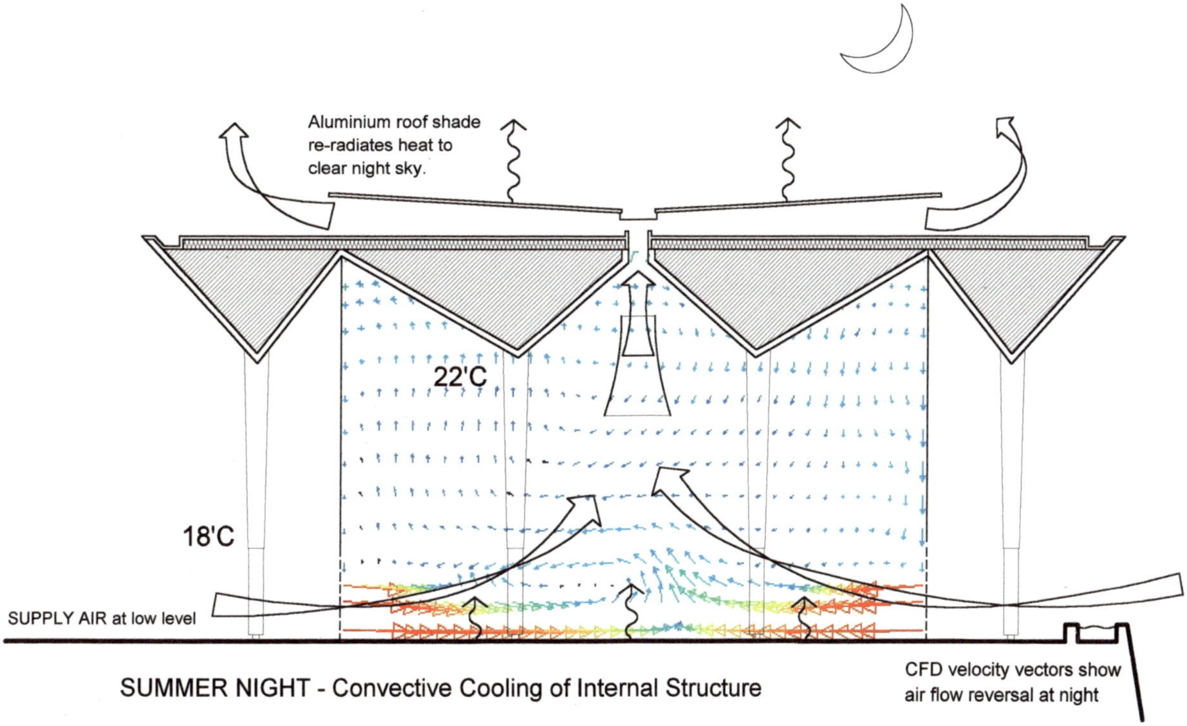

48

CASE 2 Central Shaft (Enclosed) Misting Tower

An enclosed central shaft or concourse (fig. 49) can be considered as the supply zone which serves a number of surrounding zones, and is therefore a 'multi-cell' building when considering airflow and thermal performance simulations. Case study buildings of this type include the Torrent Research Centre, Ahmedabad, and both the Seville and Catania office buildings (designed as part of the PDEC research project - Chapter 2).

The temperature in the supply zone is likely to be different from that in other zones. Also, as the height between inlet and outlets within the central shaft or concourse will vary, so too will the vent opening areas required at each level vary, if the same volume flow rate is required at each floor level. The calculation of shaft sizes and vent area opening sizes is discussed in Chapter 5.

A CENTRAL ATRIUM WITH MISTING NOZZLES CAN SUCCESSFULLY DELIVER COOL AIR TO PERIMETER SPACES TO A DEPTH OF APPROXIMATELY 12 METRES

In operation, a central atrium with misting nozzles can successfully deliver cool air to perimeter spaces to a depth of approximately 12 metres.

In Case 2, exhaust from the separate floors and individual spaces takes place via, either dedicated perimeter

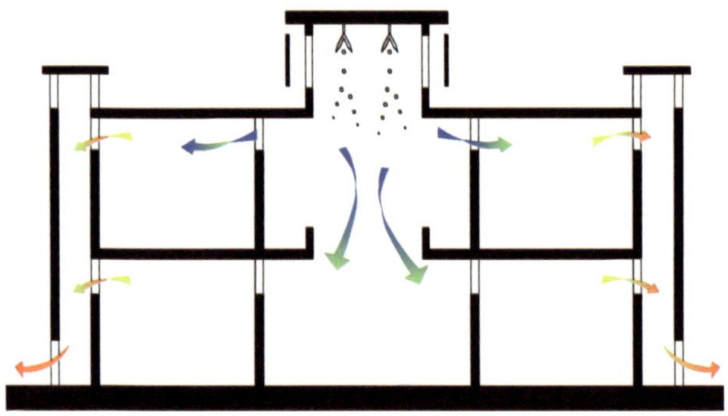

49. Case 2 Central Shaft (enclosed) diagrammatic section.

49

shafts (as in the case of the Seville office building and the Torrent Research Laboratories), or via a 'ventilated' or 'double' façade. This may be necessary in order to inhibit the impact of wind and to guarantee exhaust from the building under all conditions. It should be noted that under most conditions when PDEC is operating, the air will exit the building from the base of the exhaust towers, due to the fact that exhaust air is at a much lower temperature than ambient. When the wind is blowing, air from the top floor may exit from the top of the shafts/double façade. This can be accommodated by providing a raised parapet and vent to the top floor.

50a. Torrent Research building.
CFD Analysi of internal temperatures.

50b. Torrent Research building, Ahmedabad, India.

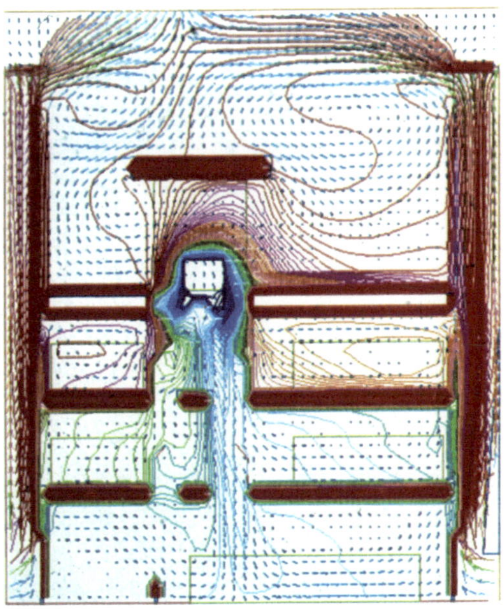
50a

50b

CASE 3
Perimeter Tower (Attached) Misting Tower

A perimeter tower attached directly to the served space or spaces can be considered a multi-cell case. (fig. 51)

The Stanford Global Ecology Centre and the auditorium in the CII Centre Bangalore, are case study buildings of this type (see Appendix A). While a central shaft has the potential to deliver air to all the spaces around it, a perimeter shaft can only deliver cool air to one side. In order to provide a sufficiently large opening from the tower into the adjacent space (to achieve the required air flow rate), it may be necessary to create a plenum along one side of the served space. Such a plenum is provided along one side of the auditorium space at the CII Centre for Excellence, Bangalore. A continuous perimeter delivery zone could form an entry/foyer transition space, where conditions are allowed to vary more widely than in the attached 'working' spaces. Alternatively, a continuous narrow buffer space could be provided in the form of a perimeter double façade. The perimeter plenum and three towers serving the auditorium of the CII Centre in Bangalore are shown in the plan and section, (fig. 53) with CFD temperature plots overlaid. (fig. 54)

Delivery from the perimeter can be successful for open plan attached spaces, but where secondary cellular spaces are included, care must be taken to ensure a satisfactory airflow path is achieved.

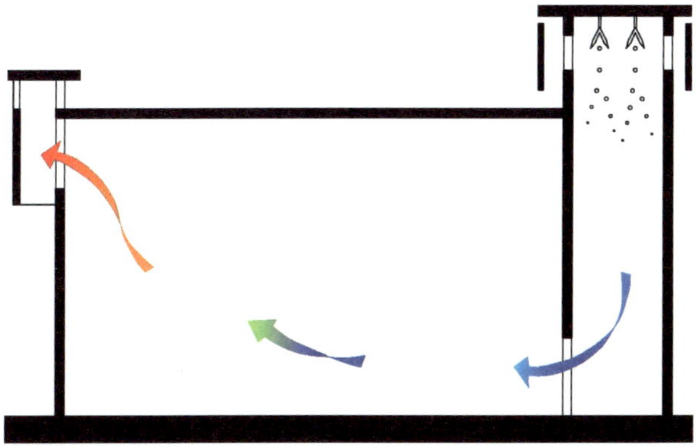

51

52

51. Case 3
Perimeter Tower (attached) diagrammatic section.

52. View of auditorium
misting towers C II Centre, Bangalore, India.

53. Section through
auditorium, C II Centre, Bangalore, India.

54. Summer day external air temperature 33°C, still air.
CFD Temperature Plot, CII Centre Bangalore.

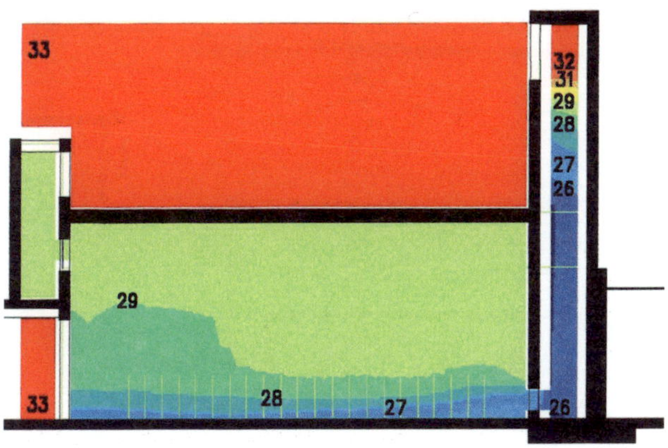

53

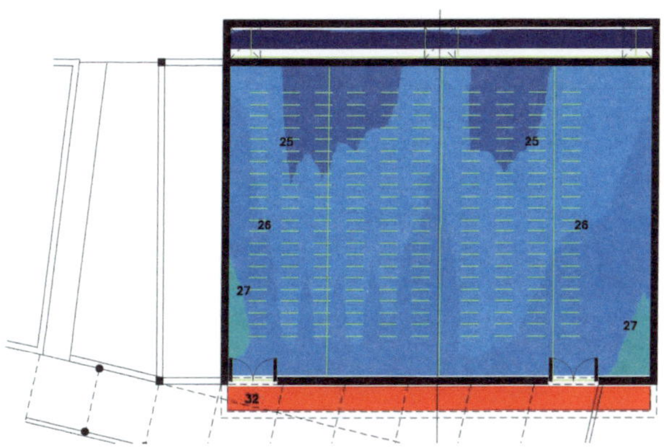

54

CASE 4
Perimeter Tower (Detached) Misting Tower

A detached perimeter tower (fig. 55) is a variant of Case 3, where for various reasons the PDEC tower cannot be located immediately adjacent to the served space. The break-out rooms at the CII Centre of Excellence are served by detached PDEC towers which deliver air below floor level into these rooms. The disadvantage of this arrangement is that the reversal of the pattern of air flow (to achieve night time convective cooling) will not take place as effectively as with the attached tower, where a vent opening can be located at high level between the room and the tower. (fig. 56)

In the case of the CII Centre in Hyderabad, India, air is evaporatively cooled at night within the tower, and the coolth stored in rocks for the following day when air is withdrawn through the rocks and delivered to the building via an air handling unit. In this case it is a hybrid system, in that pre-cooled air is supplemented by conventional chilled water within the air handling unit when additional cooling is required.

55. Case 4
Perimeter Tower (detached)diagrammatic section.

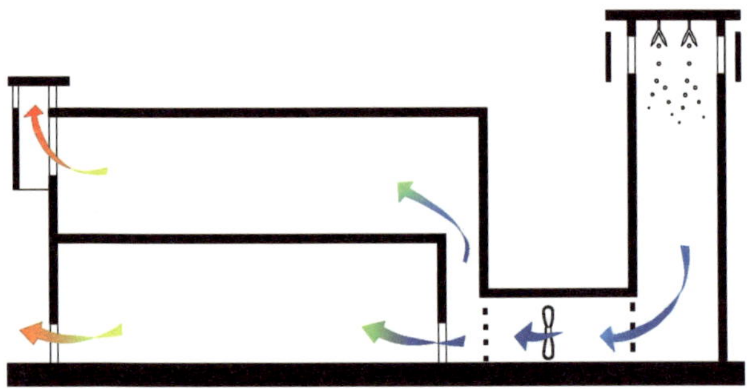

55

56

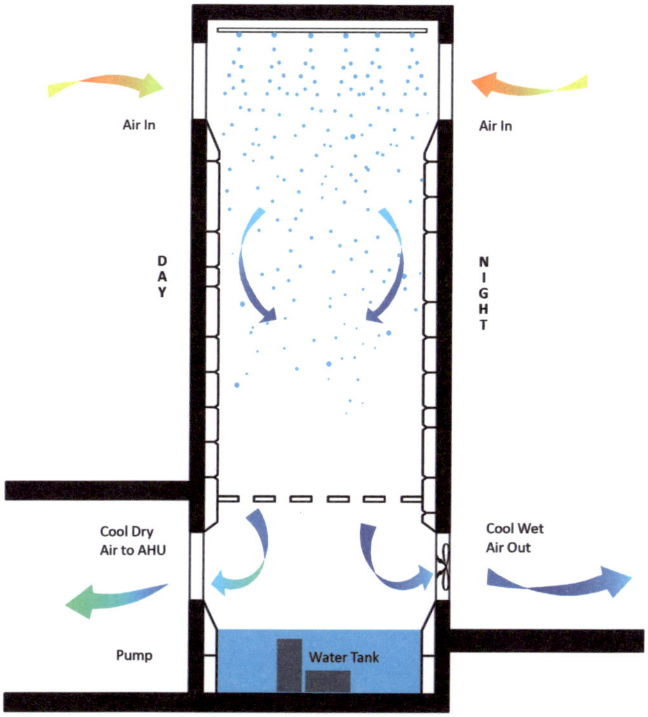

56. Misting Tower.
CII Centre, Hyderabad,
India.

57. Aerial view, CII Centre,
Hyderabad,
India.

57

Airflow Strategies
Shower Towers & Cool Towers

Airflow strategies for shower towers and cool towers are essentially the same as for misting towers, except that with shower towers and cool towers cool supply air is generally delivered at the base of the tower, whereas with misting towers cool air can be delivered throughout the height of the building. Shower towers are generally enclosed, and require a collection pool at the base of the tower. The shower tower at the Blaustein Institute in Sede Boqer, Israel, is of the 'open' atrium type (Case 1), (figs. 58 - 59) while the Interactive Learning Centre at the Dubbo Campus in Australia incorporates four shower towers of the central enclosed type (Case 2). Perimeter attached towers (Case 3) were incorporated as part of the cooling strategy for Council House 2 (CH2) in Melbourne, Australia. (figs. 60 - 61)

WITH SHOWER TOWERS AND COOL TOWERS COOL SUPPLY AIR IS GENERALLY DELIVERED AT THE BASE OF THE TOWER, WHEREAS WITH MISTING TOWERS COOL AIR CAN BE DELIVERED THROUGHOUT THE HEIGHT OF THE BUILDING

Airflow obstructions have been identified as a recurring problem in the operation of 'cool tower' buildings. Yoklic & Thompson have reported that on occasion towers have been connected to a space "with inadequate or improperly located outlets to exhaust the air". It is clear that in some buildings, while inlet and outlet to the cool tower itself has been considered, the airflow path through the building has not been adequately considered. Exhaust from the building will only be successfully achieved if the outlet free area is at least equal to the inlet free area, and if the outlets are baffled from potential positive wind pressure (this was found to be a particular problem at the Dubbo campus building). Designers must ensure that inlet or outlet openings are not obstructed and that decorative grilles, fly screens, etc, are accounted for in the calculation of effective free area.

Airflow Strategies
Porous Ceramic Panels

The air supply path containing the ceramic panels must be directly connected to the outside, which implies either a 'double' envelope or integration within a shaft or duct within the external wall, or the roof or within a standalone shaft within the building. Porous ceramic panels can be applied to cellular spaces and the shaft height will normally be limited to a single storey height. As a wall shaft therefore, an external louvered grille at high level will deliver air into the shaft, and exhaust of cool air into the room will be at low level. An equivalent area will also have to be provided to exhaust air from the room. Laboratory scale tests have established that the velocities achieved by the shaft are low (less than 0.3m/s) and may therefore be either enhanced or negated by external wind pressures.

POROUS CERAMIC PANELS CAN BE APPLIED TO CELLULAR SPACES

58. Temperatures achieved inside Blaustein Institute tower, Israel.

59. Shower tower in the Blaustein Institute, Israel.

60. Bioclimatic Section, CH2 Office Building, Melbourne.

61. Shower-Towers at CH2 Office, Melbourne.

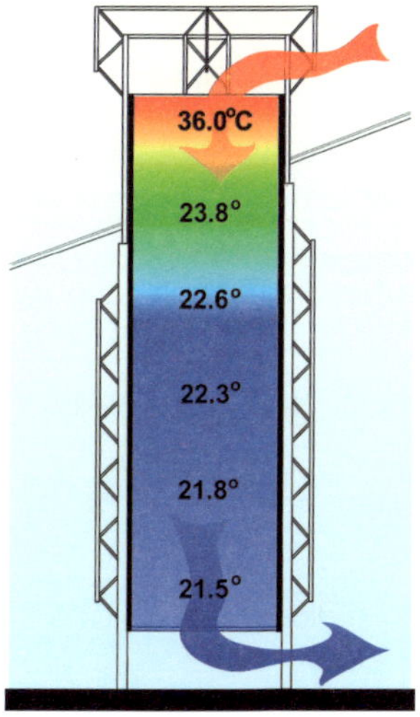

58

59

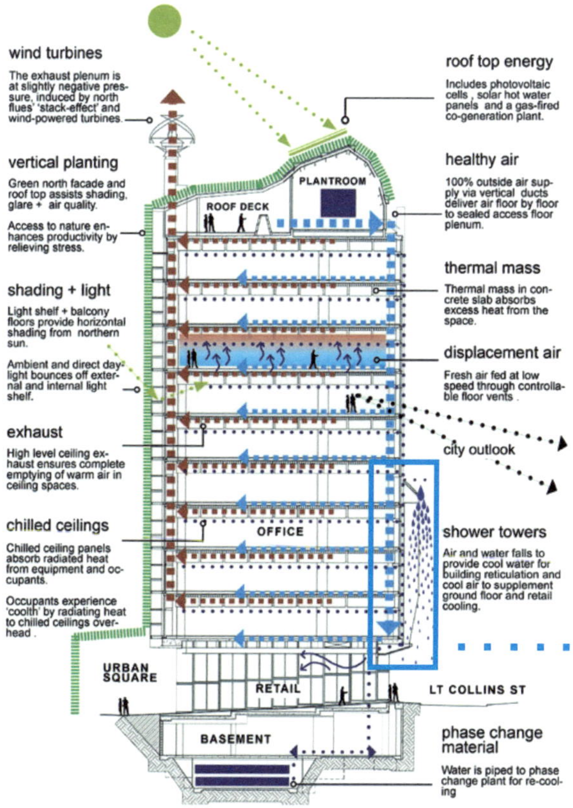

60

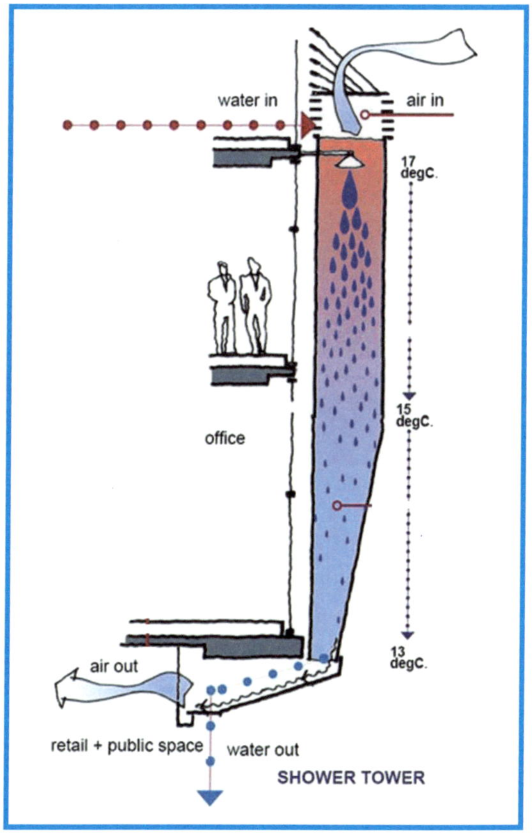

61

CHAPTER 2
INTEGRATED DESIGN

2.1 WHY INTEGRATED DESIGN?

Passive & Hybrid Cooling is not so much a technology, as an approach to design, which requires understanding and collaboration between the different members of the design team, particularly architects and engineers, to achieve what has been called 'integrated' or 'holistic' design. The manipulation of building form and fabric, the disposition of openings and the relationship between them, (normally the province of the architect) have as big an impact on the feasibility of a passive cooling strategy as the specification of pumps, vent actuators and controls (normally the province of the engineer). Great benefits arise from architects and engineers working together from the earliest stages of the design process. Given an understanding of the basic principles, passive design can provide a springboard for invention and an opportunity for innovation, based on an holistic approach by both architect and engineer. This is very elegantly demonstrated in the Tsibaou Cultural Centre in New Caledonia by Renzo Piano, and more recently in the new Centre for Sustainable Energy Technologies at the University of Nottingham campus, in Ningbo, China, designed by Mario Cucinella Architects. (fig. 62)

GREAT BENEFITS ARISE FROM ARCHITECTS AND ENGINEERS WORKING TOGETHER FROM THE EARLIEST STAGES OF THE DESIGN PROCESS

In defining a passive cooling strategy for a building, the first objective is to establish how the building form and fabric can help to minimise the cooling load. Careful design of the building envelope (thermal properties, orientation, solar access and shading, inclusion of thermal mass) can often remove the need for additional cooling in summer or heating in winter. However, depending on building type and location, summer time conditions, combined with internal heat gains, may inevitably lead to the need for some form of cooling. Also, with climate change and with the need for buildings to be robust in the face of anticipated rises in summertime temperatures, the opportunity of exploiting ambient heat sinks should be explored, in order to further reduce dependency on carbon based energy for cooling. In many situations, passive cooling techniques will meet any residual cooling requirements. Certainly passive cooling options should be considered and assessed before any decision is made to include a mechanical back-up (with attendant environmental and maintenance implications).

CAREFUL DESIGN OF THE BUILDING ENVELOPE CAN OFTEN REMOVE THE NEED FOR ADDITIONAL COOLING IN SUMMER OR HEATING IN WINTER

62

62. Centre for Sustainable Energy Technologies, University of Nottingham Campus, Ningbo, China "Mario Cucinella Architects".

2.2 INTEGRATION WITH EXISTING BUILDINGS

The influence of Urban Morphology on PDEC applicability

Studies of building typology and morphology have been undertaken as part of the process of assessing the applicability of PDEC to non-domestic buildings in southern Europe[1]. A preliminary generalised benchmark, which relates the applicability of downdraught cooling to the void/floor ratio, has been derived from the study of precedents. A method of determining the technical applicability of PDEC, based on relating the benchmark to sample areas of mainly commercial buildings in eight major cities in Spain, Portugal Italy and Greece, was applied as part of the Altener Project[2]. (Table 1)

T1

	FOOTPRINT AREA M2)	VOID AREA (M2)	SERVED AREA (M2)	RATIO VOID/FLOOR	PROP OF V/F RATIO AGAINST BENCHMARK (%)	PDEC APPLICABILITY RATING (0-1)
Seville Speculatives Office	676.0	36.0	640.0	0.056	106.5	1.00
Catania Speculatives Office	300.0	14.1	285.9	0.049	93.5	0.94
Torrent Research Centre, Ahmeadabad	376.3	38.7	337.6	0.115	217.0	1.00
Malta Stock Exchange Valletta	190.0	8.0	182.0	0.044	83.2	0.83
CII Centre of Excellence, Bangalore	317.7	6.8	311.0	0.022	41.1	0.41
PDEC Applicability Benchmark (Average of Seville & Catania)				**0.0528**	**100.0**	**1.00**

T1. PDEC Applicability Benchmark.

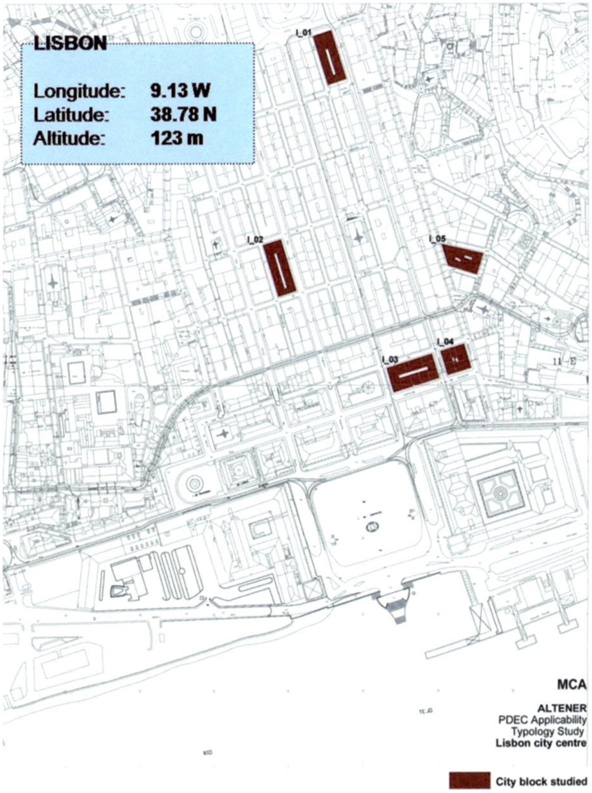

63a

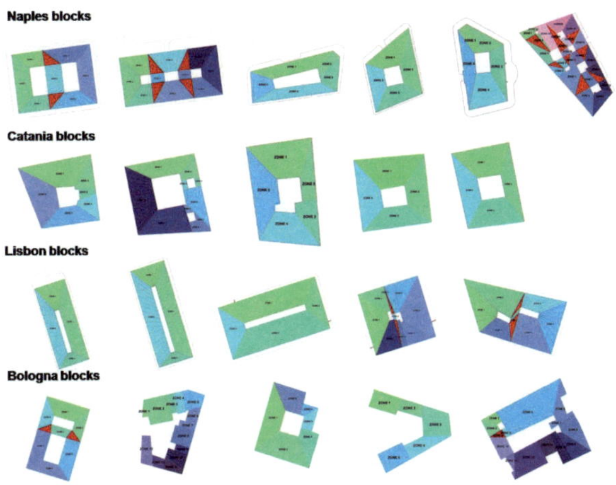

63b

63.(a) Urban morphology studies from the ALTENER research project, Lisbon.

63.(b) Urban morphology studies from the ALTENER research project city blocks identifying potential for PDEC minor and major intervention.

Existing Buildings Generic Interventions

The feasibility of integrating PDEC as part of the refurbishment of an existing building will depend on a number of factors, but a crucial determinant is whether an existing light/vent shaft or courtyard can be exploited to deliver PDEC, or whether such a shaft will have to be added to the existing structure. Studies of a number of European cities have revealed the high incidence of these features. The cooling potential of a stack relates to the achievable volume flow rate, and hence the cross-sectional area. A rough proxy for the cooling potential within a given building can then be taken as the void to floor ratio. A preliminary generalised benchmark, which relates PDEC applicability to the void/floor ratio, has been derived from the study of precedents. (figs. 63 a - b)

From Table 1, it is apparent that for existing PDEC buildings, the void: floor ratio varies between approx 0.02 and 0.11 (i.e. the voids required for PDEC represent between 2% and 11% of the floor area of the building). The more recently designed PHDC buildings have a lower void to floor ratio, which suggests that experience, and improved performance modelling, has led to a lower 'safety margin' being applied. On this assumption, a benchmark ratio of 0.05 (i.e. 5% of floor area) was chosen as the basis of a PDEC 'Applicability Rating', which could then be applied to the analysis of the existing building stock. A method of determining the technical applicability of PDEC, based on relating the benchmark to a sample area of commercial buildings in Athens, was applied as part of a 'pilot' exercise, and subsequently applied to representative areas within a number of large cities in southern Europe. The method identifies three different levels of intervention: Minor; Intermediate and Major. (figs. 64 - 65 - 66)

Minor intervention

A minor (least cost) intervention can be made when the existing building incorporates one or more light wells, with a cross-sectional area of approx 5% of the floor area. For the purposes of the Altener study, a three storey building, with an existing light well 3 - 4m across and 12m (or less) from the perimeter, was assumed. (fig. 64) The minor intervention assumes that a 3m high evaporative cooling supply air termination is added to the existing light-well, incorporating glazed baffles or glazed louvers (and possibly a glazed roof) to minimize the reduction of light penetration, and to provide an effective free area for air flow equal to the cross-sectional area of the light-well. Microniser nozzles are provided within the top of the tower. The minor intervention also assumes that the existing window openings at each floor level provide sufficient free area for the required flow rate to be achieved, and that these are manually controlled. In practice of course, many of these assumptions may not apply, depending on location, building type and other factors. However, for the purposes of the study, these assumptions then provided a basis for an assessment of both the applicability and cost of applying downdraught cooling to the existing building stock.

> A MINOR (LEAST COST) INTERVENTION CAN BE MADE WHEN THE EXISTING BUILDING INCORPORATES ONE OR MORE LIGHT WELLS

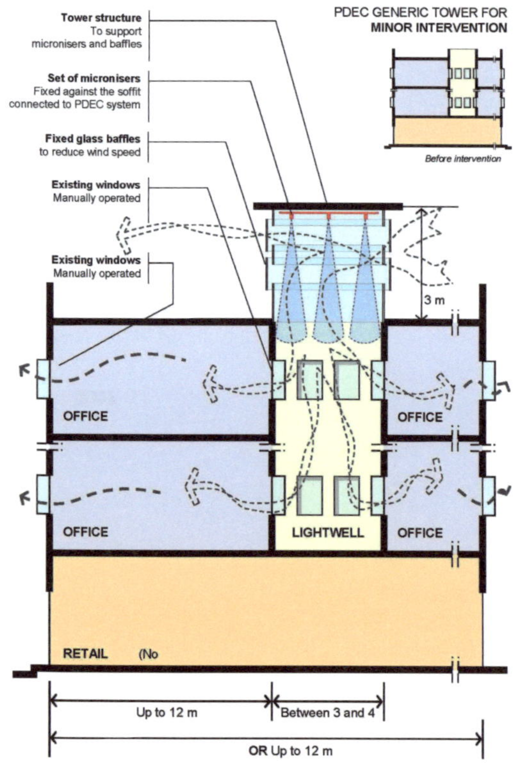

64

64. PDEC Generic Tower for Minor Intervention.

65. PDEC Generic Tower or Intermediate Intervention.

66. PDEC Generic Tower for Major Intervention.

Intermediate intervention

The intermediate level of intervention (and cost) assumes the existing building does not incorporate an existing light well, and that a new pre-fabricated lightweight downdraught cooling tower, equal in cross-sectional area to 5% of the floor area, is attached to a rear elevation. (fig. 65) It also assumes an open floor plan and a maximum distance of 12m from the tower to the opposite external wall. The new tower incorporates glazed baffles or glass louvres and nozzles as for the minor intervention. For the intermediate intervention, however, it is assumed that the perimeter exhaust incorporates motorized dampers which are controlled by the building management system. This level of intervention is very flexible and could be applied to a wide range of existing buildings, but is more expensive than the minor intervention.

THE INTERMEDIATE LEVEL OF INTERVENTION (AND COST) ASSUMES THE EXISTING BUILDING DOES NOT INCORPORATE AN EXISTING LIGHT WELL

Major intervention

During major refurbishments, involving the stripping out of existing services and extensive re-planning and re-modelling work, the creation of new light and ventilation shafts within a deep plan may be appropriate and would provide the opportunity to incorporate downdraught cooling. Such a scenario is assumed for the major intervention. (fig. 66) In this case, in addition to the new enclosed voids created within the existing plan, automated control is provided on both supply and exhaust vent openings. This is clearly the most expensive of all the options, but has the added benefit of increasing light penetration to the centre of a deep plan.

THE CREATION OF NEW LIGHT AND VENTILATION SHAFTS WITHIN A DEEP PLAN

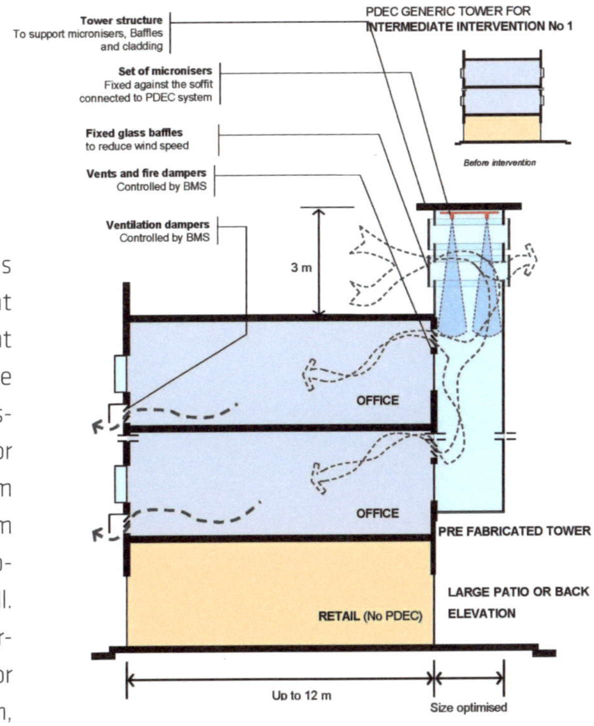

65

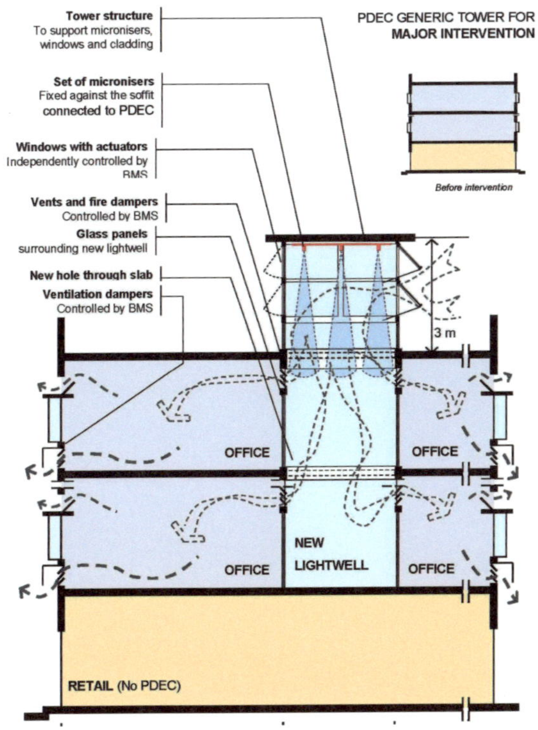

66

Applicability to existing buildings

Results of the stock analysis referred to above has revealed the proportion of the building stock which is suitable for different levels of downdraught cooling intervention. Clearly, minor and intermediate interventions are likely to be most cost effective, and will be taken up most quickly. In Spain, 35 – 46% of all buildings are capable of having either minor or intermediate PDEC interventions. This compares with 22% in Italy, 27% in Greece, and 33% in Portugal. If major interventions are included then the overall applicability of PDEC is very significant. The study indicates that, for all the cities studied, between 62% and 82% of all buildings are suitable for the application of PDEC. This indicates that the technical potential for PDEC applications is very significant. However, local micro-climate characteristics, (fig. 67) market barriers, and the cost of the systems, will influence the final potential market penetration.

67. Map of PHDC applicability (Tdb – Twb) in Europe.

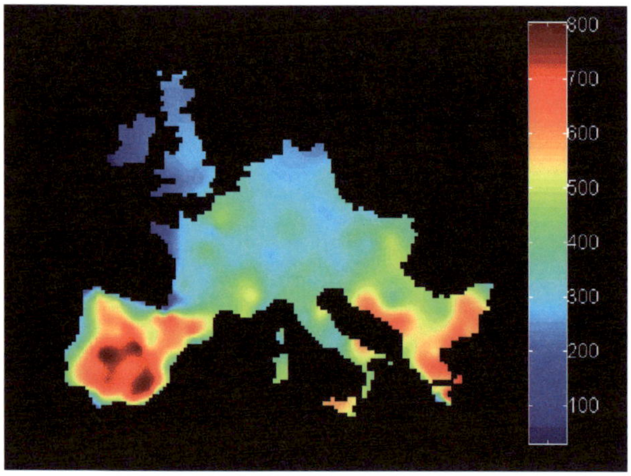

67

2.3 INTEGRATION IN NEW BUILDINGS

Downdraught Cooling has been adopted as a design strategy to promote low carbon cooling in over 100 buildings world-wide over the last 10 – 15 years. Ten of these buildings have been the subject of post-occupancy performance evaluation, and are described in detail in the attached CD. There are also a number of un-built projects which have contributed to our knowledge of this approach to cooling. Two un-built projects for new office buildings for locations in Seville and Catania in southern Europe are described here as they were the subject of detailed performance and cost analysis as part of the EC funded PDEC project[3]. (fig. 68)

68. PDEC
life cycle cost comparison
with 'comfort cooling'.

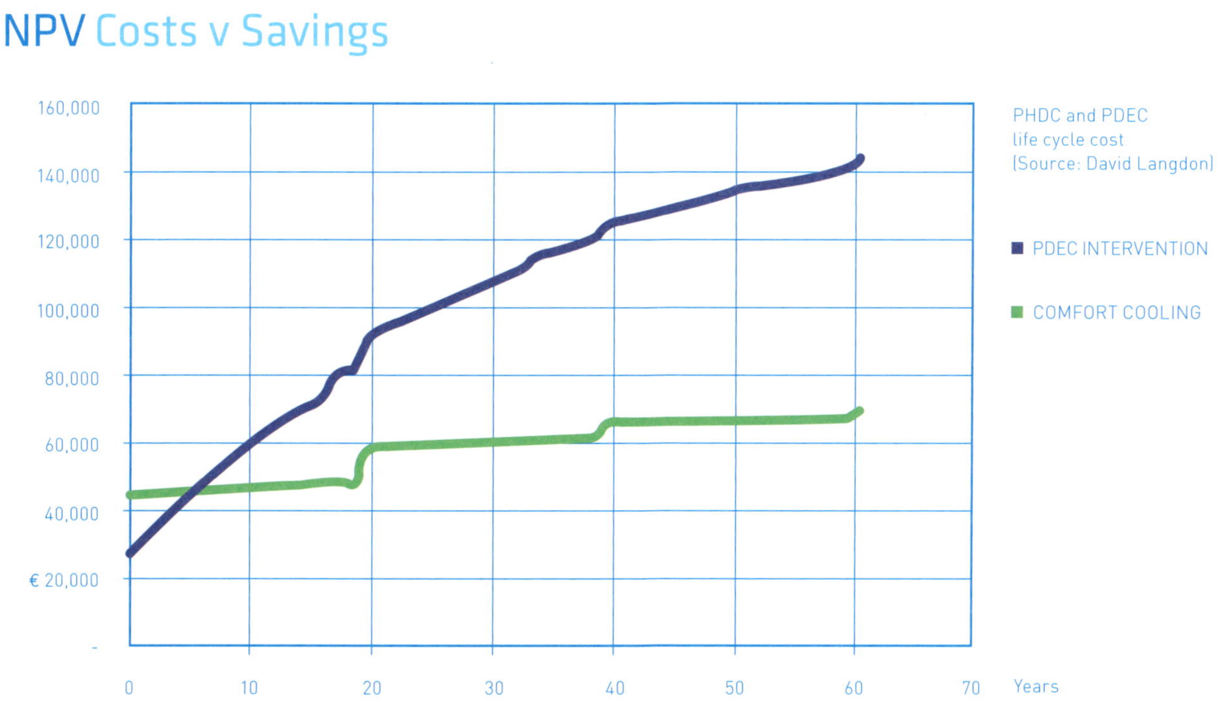

68

Office Building Seville, Spain

The design of this commercial office building in the busy heart of Seville, by Brian Ford Associates, dealt simultaneously with the commercial requirements of the brief, the constraints of the urban site and geometrical implications of the downdraught cooling strategy. (figs. 69 - 70) The design[4] attempts to achieve an essentially 'deep plan' office floor plate (overall width 27m) incorporating substantial (6m wide) light wells, which also serve as the 'transitional space' through which evaporatively cooled air is distributed to each floor. (fig. 70) The open shafts in the centre of the building are surrounded by open plan offices, with cellular offices and meeting rooms around the perimeter. (fig. 73) The overall floor plate width of 27 metres is extremely ambitious, but it was found through CFD and dynamic thermal analysis[5] that this geometry can promote adequate air movement and satisfactory temperature distribution. Part of the design task was to establish a threshold for the depth of plan and the degree of cellularisation which can be achieved using passive downdraught evaporative cooling.

IN ORDER TO DEVELOP A VIABLE COOLING STRATEGY IT IS VITAL TO ENSURE THAT SOLAR HEAT GAIN IS REDUCED TO THE MINIMUM

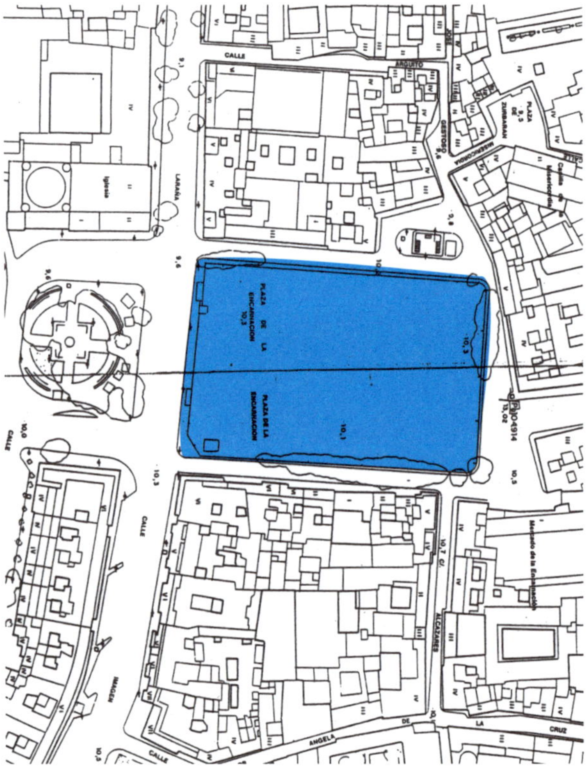

69

In order to develop a viable cooling strategy it is of course vital to ensure that solar heat gain is reduced to the minimum. Solar access is controlled by combined sunshades and external light-shelves to south facing windows. They are designed to maximise reflected light

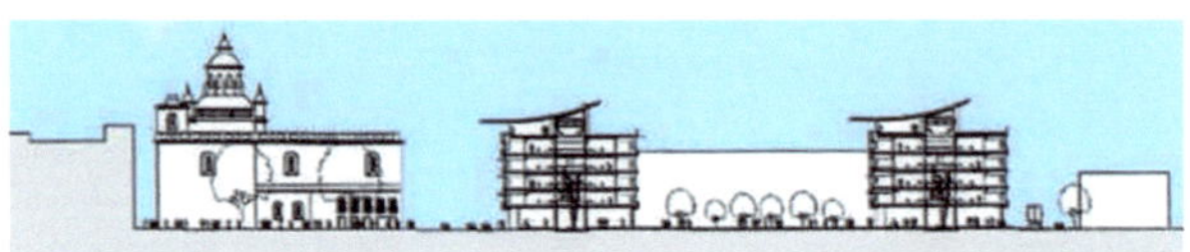

70

69. Office site and immediate context Seville, Spain.

70. Site section showing proposed office buildings, Seville, Spain.

71

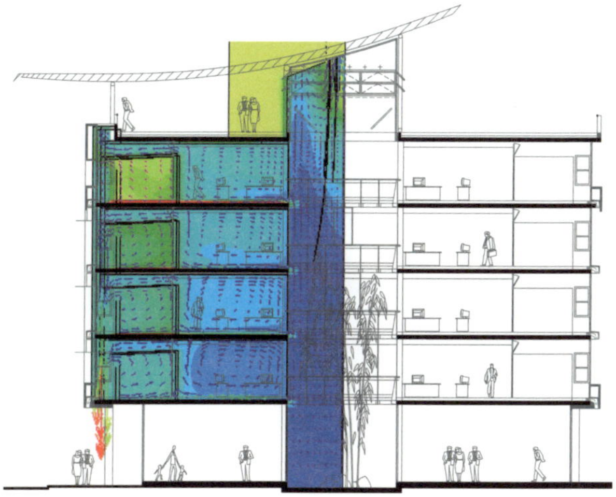

72

71. Sectional model of office building, Seville.

72. CFD vector + temperature plots overlay on office section, Seville.

into the back of the offices. The glazing at the top of the open light-well is protected with fixed louvres adjusted to exclude direct solar in summer while allowing sunlight in winter. (fig. 71)

(figs. 71 - 72) Offices on the south (street) side of the building must be protected from noise and pollution and so it was proposed to exhaust air via shafts which allow stale air to either fall (in still air conditions) or rise (in windy conditions). Since the prevailing wind is also from the south-west, this approach avoids the potential problem of positive pressure on exhaust air vents. On the courtyard side, air is exhausted via casement and clerestory glazed vents. Exhaust air from open plan offices passes through the bulkhead duct located in each office, and out via automatically controlled top hung ventilators. (figs. 73)

In the summer the volume of supply air is controlled by glazed motorised louvres which respond to wind direction and ambient temperature. Distribution of air from core to perimeter employs two routes: either through a dedicated bulkhead direct to vertical shafts at the perimeter or via perimeter cellular offices. Performance simulation using computational fluid dynamics confirmed the viability of this strategy.

Night-time convective cooling coupled with the mass of the building is part of the strategy for PDEC to be successful in the climate of Seville. This is achieved by encouraging flow reversal in both still and windy conditions. Control of louvres at high level is necessary to ensure negative pressure at the outlet. Winter daytime ventilation is achieved via trickle vents located in the soffit of the cantilevered slab at each level. This detail provides protection from the prevailing wind and allows pre-heating if required. Stale air is exhausted via the light-wells and high level vent. Motorised louvres would be controlled to block the prevailing wind in winter in order to encourage air to exhaust on the leeward side.

Performance analysis undertaken at De Montfort

University[6-7] revealed that for the Seville Office Building, the wet bulb temperature is sometimes too high to rely on evaporative cooling all the time. Mixed mode operation is therefore proposed for this office building to maintain internal air temperatures below 26.°C. Nevertheless, simulation results indicate energy consumption approximately 75% below the air conditioned reference building. Also, overall building costs have been estimated to be 6% below the air conditioned reference building. This speculative office building in Seville was created as a design exercise as part of the research project into the applicability of PDEC in Southern Europe, and has yet to be realized in practice.

SIMULATION RESULTS INDICATE ENERGY CONSUMPTION APPROXIMATELY 75% BELOW THE AIR CONDITIONED REFERENCE BUILDING

73. Plan of Commercial Office Building, Seville. showing atria (yellow) & perimeter shafts (red).

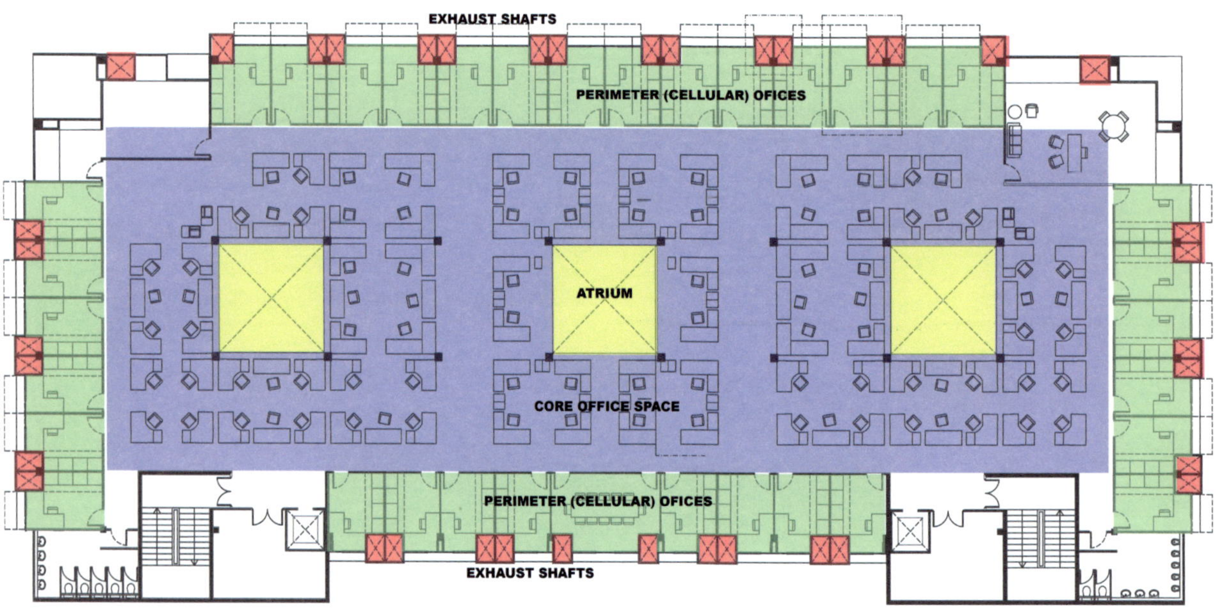

73

Office Building Catania, Sicily

In parallel with the design of the Seville office building, Mario Cucinella Architects designed a new office building for Catania, Sicily, as part of the same 'PDEC' research project. This project followed the construction and testing of a full scale prototype experimental building located at the Conphoebus Research Institute in Catania. The climate in Catania is hot and dry in the summer months, although the wet bulb occasionally goes above 24°C, reducing the time during the summer when evaporative cooling is effective. The performance of the experimental building was extensively monitored and the data collected helped to inform the design of the new office building.

A brief was developed for an owner-occupied office building that should respond to standard market demands for spatial and functional requirements. The building is four floors high and houses offices, a conference centre and has a floor area of 4000m2. The approach to the building design was orientated towards environmentally friendly architecture that would demonstrate architectural quality, use building materials and approaches that are easy to maintain, are not harmful to the environment and above all enhance the technical performance of the downdraught evaporative cooling system.

The initial design phase identified the zones of the building to be served by downdraught evaporative cooling and developed the mechanism to move the cooled air to the spaces. An initial proposal with a large central atrium presented difficulties in terms of efficiency and control of the distribution of the air between the atrium and the office spaces. A large volume of air needed to be cooled even at times when the cooling demand was low or varied between different office spaces. It was difficult to control the movement of the air into the lower floor spaces and to achieve uniform air speeds across the office spaces. The atrium typology was also inefficient in terms of the construction costs and usable floor area thus decreasing the overall sustainability of the design.

These observations caused the designers to reconsider their approach to the design and return to the initial research question – 'is there a 'PDEC' building – a building whose form and expression is derived from the passive downdraught evaporative cooling phenomenon?' – and to re-examine what was learned through the design of the experimental building. (fig. 74)

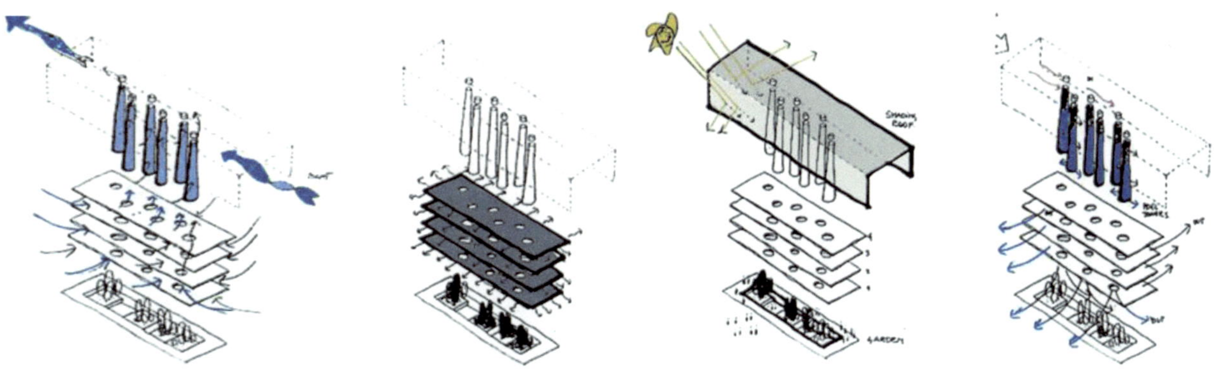

74

74. PDEC office building, Catania Environmental strategy diagrams.

THE TOWERS WERE DEVELOPED AS AN ARCHITECTURAL AND ENVIRONMENTAL ELEMENT PROVIDING NOT ONLY COOLING BUT STRUCTURE, VENTILATION AND DAYLIGHT

When the micronisers spray water into hot dry air, a plume of cool downward moving air is created. This phenomenon was exploited in the experimental building and became the starting point for the revised new building design. Instead of having a central atrium and a volume of cooled air, the building has a number of 'PDEC' towers crossing the building vertically.

This system has a number of advantages over the atrium reservoir approach:

- The PDEC towers combine and exploit the concepts of evaporative cooling with the downdraught effect.
- It is possible to cool different areas of the building at different times to satisfy its real needs.
- Each individual tower can be defined to provide the cooling power required for each area immediate to the tower. The EVAPORATIVE COOLING system in each tower can be switched on or off as needed, cooling only enough air as required by the zone served by that tower. This means that the amount of cooling can be controlled to respond to the varying demand in each part of the building.
- The building management system can be more accurate thanks to the increased sensitivity of the system.
- Natural light can be distributed allowing an increase in the floor depth without reducing comfort. (fig. 75)

The towers were developed as an architectural and environmental element by providing not only cooling but structure, ventilation and daylight, functions that in office buildings are generally split into separate service components. The building is considered as a natural organism that reacts and adapts to external changes in a sensitive way. This new strategy enables greater sensitivity to spatial cooling demands and the simpler shape and deeper plan engenders significant reductions in construction costs and overall energy consumption.

The design development concentrated on evaluating the efficiency of the PDEC cooling towers, defining their size, position and number, the depth of the plan, the position of the openings, the structural scheme, the facades, the contribution of the towers to daylight levels in the offices and the office layout.

The PDEC Tower: Building Performance Assessment

Early simulations of the tower performance by the Building Performance Assessment team confirmed the potential of the towers in terms of cooling, comfort and energy efficiency. A comparison of the preliminary design (atrium) and final design (towers) showed a significant reduction of both peak cooling (25%) and cooling requirements (between 30% and 50% lower). (fig. 76)

The height of the tower and different alternatives of opening were evaluated using CFD simulations. The objective was to find a combination able to distribute the air flow rates in a consistent way to the cooling requirements of the spaces. It was found that the openings in the tower with a low aspect ratio (width over height) close to 0.1, reduced problems such as the direction of the air flow leaving the opening into the building, the differences existing between the top floors of the building and the bottom floors and the existence of reverse flow conditions. Increasing the height of the tower further reduced the difference between the opening areas required at the different floors.

THE HEIGHT OF THE TOWER AND DIFFERENT ALTERNATIVES OF OPENING WERE EVALUATED USING CFD SIMULATIONS

Layout influences on internal conditions were simulated with FLUENT 4.4. The simulations gave clues as to how the towers could be refined to improve the uniformity of air velocity and temperature distribution throughout the office spaces.

The most efficient configuration was found to be:

- Exclusion of the effects of the wind so the air entering the towers is considered to be a zero velocity. A tower 5m high above the roof level allows complete evaporation

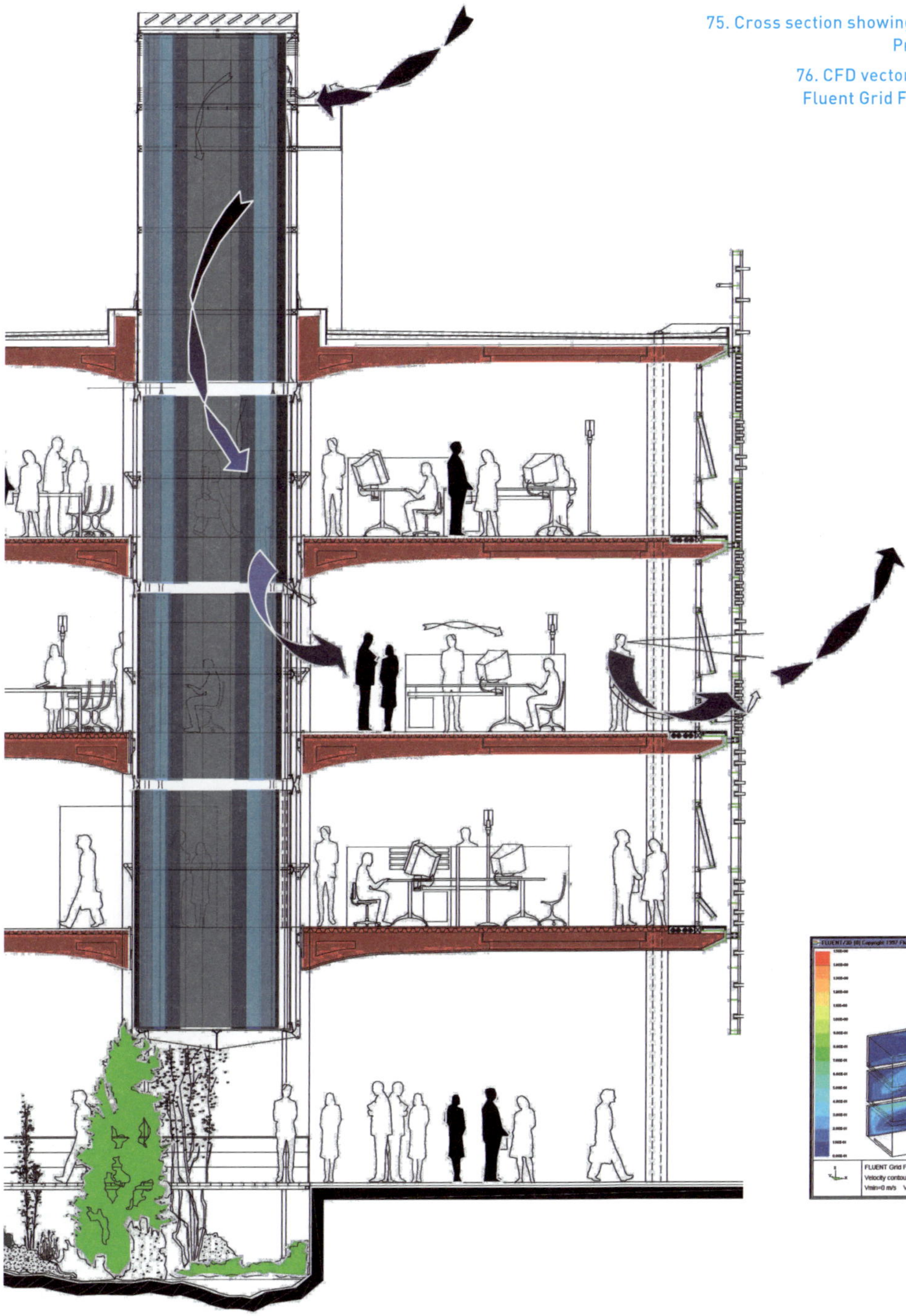

75

75. Cross section showing downdraught cooling strategy. Proposed office building, Catania.

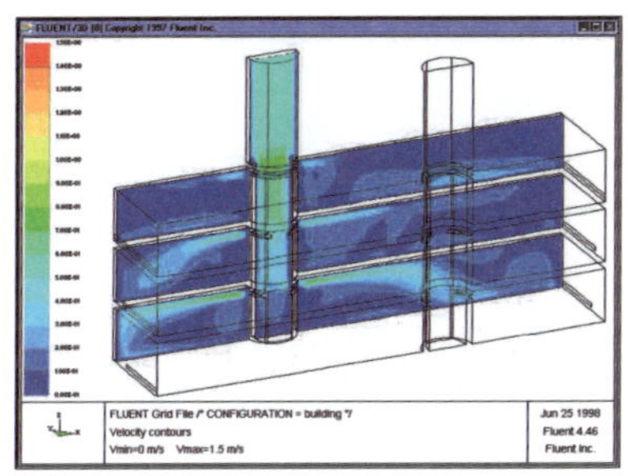

76

76. CFD vector plot for Office BuildingCatania. Fluent Grid File /* Configuration = building */

of the water from the micronisers before the air enters the offices.

• Two towers with high level inlets and low-level outlets gave the best results in terms of temperature and air velocity distribution throughout the spaces.

• A layout with towers placed alternatively throughout the building floors at 6m centres and 27m wall to wall.

• A tower diameter of 3m.

• 200mm high uniform openings around the top of each tower.

Energy Savings

A strategy for extra cooling and heating was defined to cover all the seasons and to establish the energy consumption over twelve months. The figures resulting from the simulation were encouraging as the overall energy consumption of this building was estimated to be 15% of that of a conventional building of the same size. Simulations also showed that the temperature and air speed was reasonably uniform throughout the office spaces.

77

77. Sectional model for proposed Office Building Catania, Italy.

Building Structure

As part of the development of the tower design as an integrated multifunctional component of the building, a structural scheme was devised that is generated by the geometry of the PDEC towers.

The floor slabs sit on structural rings that are supported by four concrete pillars at each tower. The rings are horizontally braced between each other and are connected to the pillars by curved concrete ribs. Along the façade a series of single columns support the perimeter. This scheme concentrates all the structural elements around the cooling towers and the façade and leaves the plan free from columns. The glazed internal façade of the tower is supported from this structural system. Two service blocks at the ends of the building provide structural stiffness to the ensemble. This type of structure emphasizes the geometry of the building and enhances the organic continuity between the different components.

Office Layout

The introduction of the PDEC towers into the building defines the spaces and creates areas that reflect changing spatial needs in today's workplace. The layout design is flexible so as to include cellular office spaces, flexible meeting area and open plan desk areas.

The final office layout is based on the following criteria:

- The deep-plan solution (27m) allows better use of the space and more flexibility. The PDEC towers passively provide levels of thermal and visual comfort not usually found in deep plan buildings.
- The PDEC tower creates a modular unit providing structure, lighting, ventilation, and cooling. This element accommodates most of the building services, therefore

78. Long section for proposed office building, Catania.

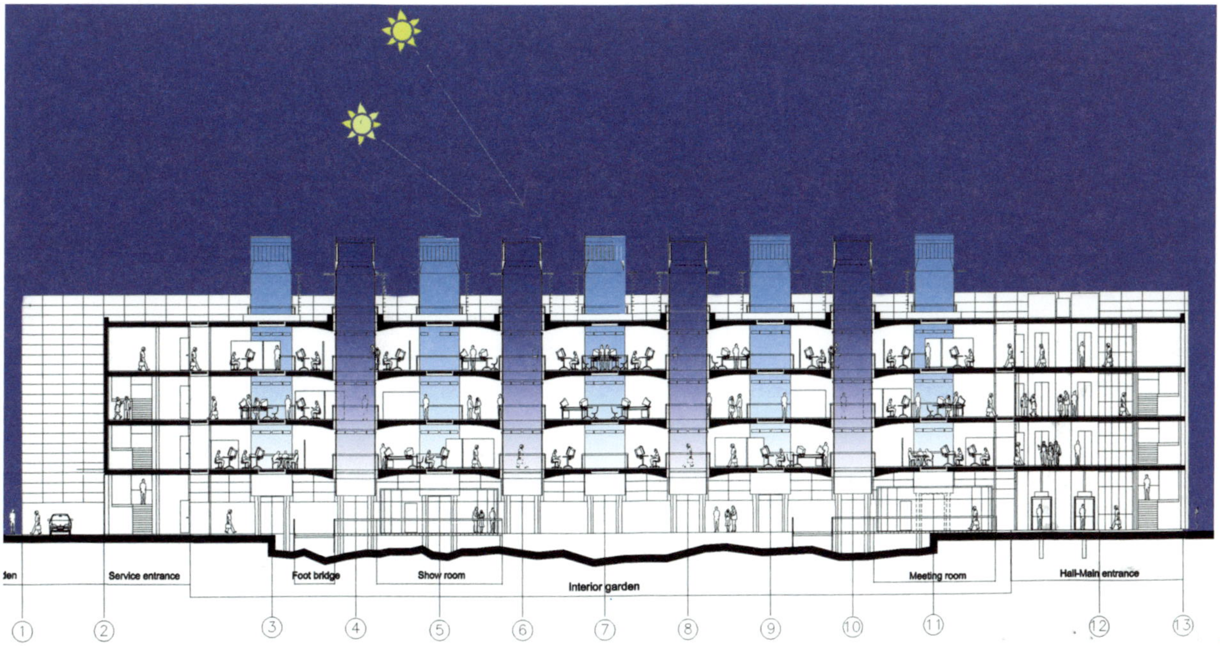

78

increasing the usable surface area.

- Circulation and shared areas are concentrated in the central part of the plan between the towers.
- An open plan distribution is used, with the permanent working desks along the facades and the more flexible ones (hot desking, desk sharing), on the central axis.

Tower Elements

The tower is composed of the "body", that crosses all the floors, and the "head" that rises over the roof. The head acts as the air inlet and as a light catcher and diffuser. It also contains the ring of micronisers and the evaporative zone (4,2m high) The tower envelope above the roof of the building is made from an insulated metal sheet on a metallic frame with a mirrored internal surface and the top is fully glazed to let light in. The air inlet is designed so as to control the wind effect and avoid turbulence inside the tower.

The body distributes air and light to the offices. It is a glazed 3m diameter cylinder with a 200mm opening around its perimeter close to the ceiling at each floor. The structural glass body provide maximum transparency at the lower floors and has reflective strips at the higher floors. These strips are made with a dotted serigraphy that is 50% reflecting and 50% transparent.

A GLAZED 3M DIAMETER CYLINDER WITH A 200MM OPENING AROUND ITS PERIMETER CLOSE TO THE CEILING AT EACH FLOOR

The External Facades

The facades are composed of a double layer. The internal layer is single glazed with aluminium mullions. There are 40mm openings for the air outlet at the bottom and openings for cross ventilation (free running situation) in the middle. The external layer is the "massive" part of the façade. It is composed of stone elements (tufa) that provide shading to the internal layer. The stone shading elements are 50 cm long and 5cm thick with a variable width depending on the need for shading. They are supported on a metallic frame composed of steel plates. The shading elements are positioned on horizontal lines, creating a texture with a variable density: more dense on the top floor and less dense on the lower floor, in order to balance the natural light provided by the towers. The shading is spaced so as to provide maximum shading in summer and allow useful solar gain in winter.

THE TOWERS ARE SUCCESSFUL IN REDUCING THE NEED FOR ARTIFICIAL LIGHT

Daylighting studies

In order to evaluate the contribution of the PDEC tower to the daylight levels in the office space, daylighting studies were undertaken at Lausanne Polytechnic's LESO laboratory in Switzerland.

A 1:50 scale model was produced for testing under an artificial sky. The model consisted of two modules of the building and interchangeable tower types with different amounts of transparency and reflectance. An experimental protocol was prepared to determine the positions of the sensors in the model and the different tests to be undertaken. Four tower types were tested in combination with two types of façade. Eight sensors were placed inside the model and one on the roof and measurements were taken under a standard CIE overcast sky. The daylight factor was calculated from the sensors and the results of the different tests compared.

From the analysis of the results it was found that the towers did contribute to the light levels in the centre of the building. The most efficient solution, a tower with 100% reflectance at the roof level, increased the daylight factor by 10% compared to the same model without the towers. The measurements have demonstrated that the towers are successful in reducing the need for artificial light, which means a reduction of energy load and a better standard of visual comfort in the building.

79

Conclusions

The desire in this project was to develop a methodology for architectural research by exploring the potential of a scientific phenomenon through design. The intention was not only to achieve a viable working passive downdraught evaporative cooling system but to develop an architectural language that could bridge the gap between historical and contemporary design. (fig. 79)

79. PDEC office building Catania
glazed towers
in the office space.

CHAPTER 3
CASE STUDIES

3.1 OVERVIEW

In this chapter we look at the performance and occupant responses to a number of buildings in different parts of the world which have incorporated PHDC. Ten buildings, from a total of over 100 worldwide, which have applied this approach to cooling, have been investigated in detail. The sample includes three in India, four in the USA, one in the Middle East, one in Southern Europe and one in northern Europe. All the buildings surveyed were non-domestic. While the sample is small, some general observations can be made which may be of more widespread interest and application:

1. The survey of ten case study buildings has shown that PDEC can provide significant energy savings, significant life-cycle cost benefits, and high levels of occupant satisfaction. Some of the buildings have demonstrated all of these characteristics. Some have demonstrated small energy savings and a relatively low level of occupant satisfaction. All the buildings surveyed were non-domestic.

2. The ten buildings surveyed included buildings only cooled using PDEC, buildings using both PDEC and cooling coils, and buildings using only cooling coils. Generally, hybrid buildings are more complex and involve a greater management challenge.

3. Misting nozzles have been shown to be generally the most efficient (in terms of water use), flexible, easy to control and low maintenance technique for the creation of passive downdraught evaporative cooling.

TEN BUILDINGS, FROM A TOTAL OF OVER 100 WORLDWIDE, WHICH HAVE APPLIED THIS APPROACH TO COOLING, HAVE BEEN INVESTIGATED IN DETAIL

4. Cellulose and other porous panels have been used successfully, but tend to be inefficient in water use, and difficult to control.

5. Cooling coils have been shown to be effective in delivering downdraught cooling (and saving fan energy) in humid conditions.

THE SURVEY OF TEN CASE STUDY BUILDINGS HAS SHOWN THAT PDEC CAN PROVIDE SIGNIFICANT ENERGY SAVINGS, SIGNIFICANT LIFE-CYCLE COST BENEFITS, AND HIGH LEVELS OF OCCUPANT SATISFACTION

6. Maintenance of PDEC systems is regarded as far less onerous than for conventional air-conditioning. However, hybrid systems require expertise to run them successfully, as they are more complex and building managers require training.

7. Occupant surveys reveal that PHDC buildings in India are more favourably perceived by their occupants than those in the USA. This suggests that culture as well as climate influences expectations, although the reasons for these differences are likely to be multi-faceted.

8. There is significant variation across the sample in terms of occupant satisfaction. Although four out of ten had internal environments which were perceived as being significantly better than benchmark values, a similar number had results which were significantly below benchmark.

Three of the ten buildings for which post-occupancy surveys have been carried out are presented in this chapter. The other seven buildings are described in detail in the appendix which can be found on the attached CD.

3.2 FEDERAL COURTHOUSE, PHOENIX, ARIZONA, USA

Local Context

In the summer of 2007 Phoenix, capital city of Arizona, set a record of 32 consecutive days with temperatures above 43°C. Located on the north margin of the Sonoran desert (Lat. 33° 26' N, Long.112° 04' W), the town experiences very hot and dry summers (40.5°C mean maximum for August) with peak temperatures of 50°C and mild winters (4.1°C mean minimum in January). (fig. 80) The annual mean minimum temperature is 14°C and night-time temperatures drop considerably lower than the daytime, especially in winter, spring and autumn. The relative humidity in summer is also very low with averages of 10-20% which, together with the prevailing easterly winds, increases the outdoor comfort threshold. (fig. 81) Phoenix and its surrounding area benefit from 325 days of annual sunshine but receive only 194mm of rain per year[1], making this area ideal for the application of solar energy technologies but also vulnerable to drought. Water availability is, in fact, one of the major concerns in the south-western USA.

PHOENIX AND ITS SURROUNDING AREA BENEFIT FROM 325 DAYS OF ANNUAL SUNSHINE BUT RECEIVE ONLY 194 MM OF RAIN PER YEAR

80. Monthly diurnal averages for Phoenix, plotted with Climate Consultant 4 .

81. Psychrometric chart for Phoenix, plotted with Climate Consultant 4.

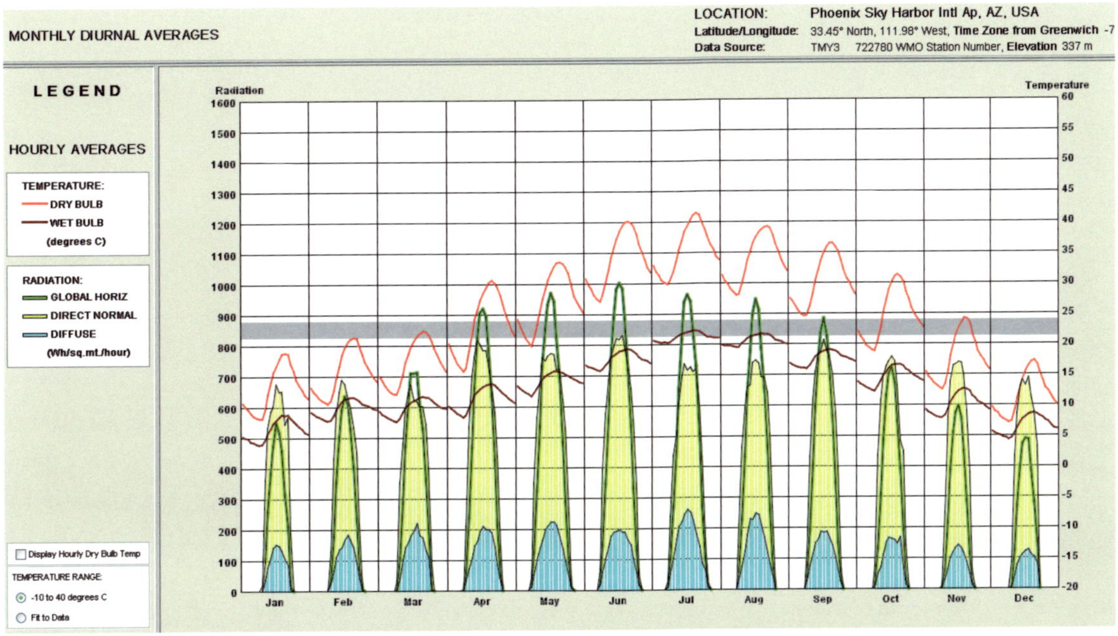
MONTHLY DIURNAL AVERAGES
LOCATION: Phoenix Sky Harbor Intl Ap, AZ, USA
Latitude/Longitude: 33.45° North, 111.98° West, Time Zone from Greenwich -7
Data Source: TMY3 722780 WMO Station Number, Elevation 337 m
LEGEND
HOURLY AVERAGES
TEMPERATURE:
DRY BULB
WET BULB
(degrees C)
RADIATION:
GLOBAL HORIZ
DIRECT NORMAL
DIFFUSE
(Wh/sq.mt./hour)
Display Hourly Dry Bulb Temp
TEMPERATURE RANGE:
-10 to 40 degrees C
Fit to Data
Radiation
Temperature
Jan
Feb
Mar
Apr
May
Jun
Jul
Aug
Sep
Oct
Nov
Dec

80

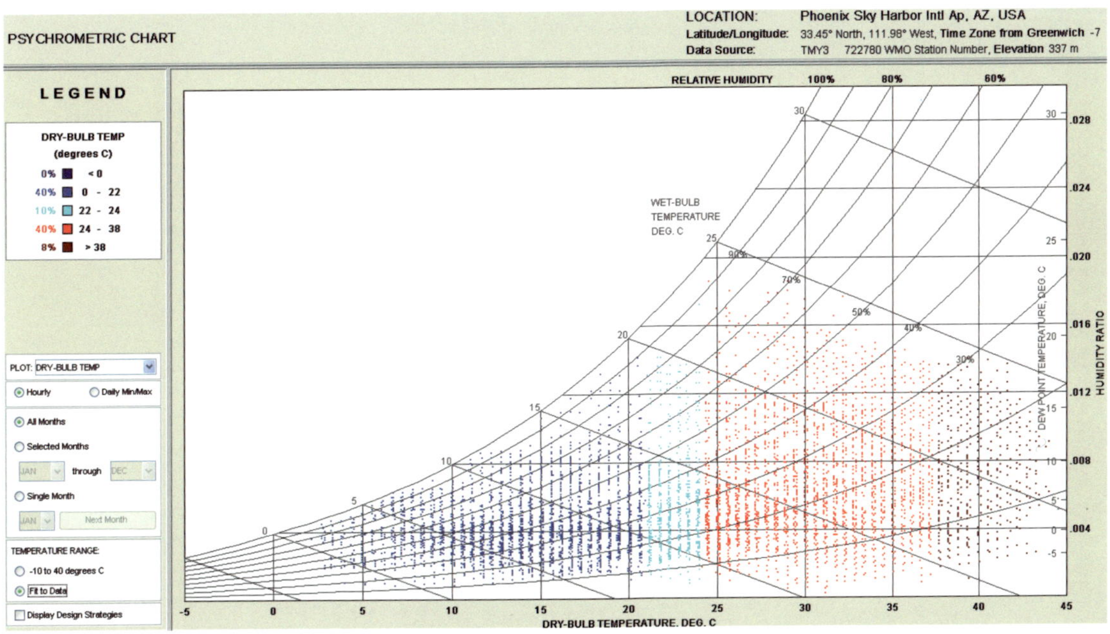
PSYCHROMETRIC CHART
LOCATION: Phoenix Sky Harbor Intl Ap, AZ, USA
Latitude/Longitude: 33.45° North, 111.98° West, Time Zone from Greenwich -7
Data Source: TMY3 722780 WMO Station Number, Elevation 337 m
LEGEND
DRY-BULB TEMP
(degrees C)
0% < 0
40% 0 - 22
10% 22 - 24
40% 24 - 38
8% > 38
PLOT: DRY-BULB TEMP
Hourly
Daily Min/Max
All Months
Selected Months
JAN through DEC
Single Month
JAN Next Month
TEMPERATURE RANGE:
-10 to 40 degrees C
Fit to Data
Display Design Strategies
RELATIVE HUMIDITY 100% 80% 60%
WET-BULB
TEMPERATURE
DEG. C
DEW POINT TEMPERATURE, DEG. C
HUMIDITY RATIO
DRY-BULB TEMPERATURE, DEG. C

81

The building and the overall strategy

Phoenix's urban form is similar to that of many new south western cities in the USA. The very low density suburbs give way to the high density centre downtown. The road infrastructure, however, stays the same and the multi-lane carriageways continue deep into the city centre. In this context, located on the west margin of downtown Phoenix, the 46,500m2 six storey courthouse sits on two urban blocks. The building is oriented north south with the main entrance on the east side, preceded by a paved plaza. It consists of two volumes: a 6 storey office and courtroom block and a larger 6 storey atrium space. (fig. 82)

82. Federal Courthouse. Interior view.

83. Diagrammatic plan showing main building zones Federal Courthouse, Phoenix, Arizona.

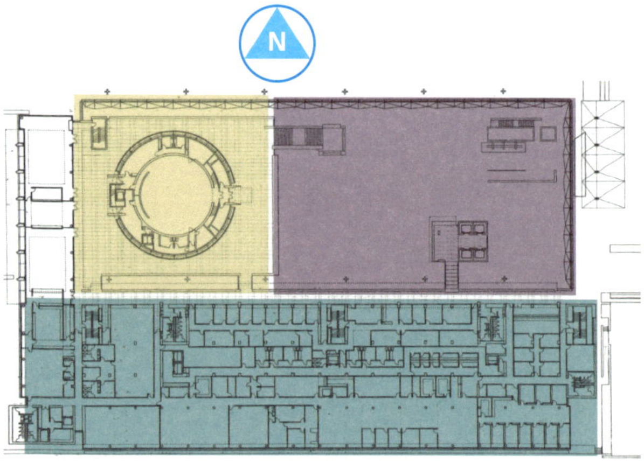

83

82

THE ATRIUM SPACE USES PASSIVE DOWNDRAUGHT EVAPORATIVE COOLING (PDEC)

Despite the large glazed area, this is an inward looking building concentrating most of the offices and courtrooms on the south side. These areas open into the vast 107x46m north atrium via the open access balconies. (fig. 82) On the west side, the atrium houses the 3-storey cylindrical special proceedings courtroom, which is located at podium level The atrium incorporates ventilation openings at low level on the east side and at high level on the north side and roof. The atrium's glazing is either shaded clear double glazing on the north elevation or white ceramic fritted low-e glass on the east and west sides.

Design intentions and architect's philosophy

The design team's primary intention was to create a building that transposed an outdoor 'public' space where more commonly the car-park is located, into an internal transitional space between the extreme outdoor conditions and the controlled courtroom and office environment. The space acts as an imposing 'lobby' and public plaza where the citizens of Phoenix can congregate in a way they would not be able to do outside. The atrium therefore shelters them from the harsh desert climate, while allowing diffuse daylight and sufficient air movement to mitigate the impact of the intense solar radiation and high temperatures.

Environmental strategy

The building generally uses conventional air-conditioning systems to provide cooling in summer and heating in winter to the offices and courtrooms, but the atrium space, uses Passive Downdraught Evaporative Cooling (PDEC). This produces a 75% saving of the total air-conditioning costs electrical energy, and 5M$ in the equivalent capital cost of air-conditioning. The system is operated via the Building Management System (BMS) and is controlled by temperature set-points.

One of the crucial aspects of the summer strategy is to reduce solar gains from the atrium's west and east elevations and the glazed roof, whilst maximising the daylighting. This is achieved with an extensive use of white horizontal external shading, Daylighting is overall sufficient in the atrium but the use of artificial lighting becomes unavoidable as we travel further away from the atrium corridors, creating a very typical deep plan office environment. Artificial lighting is operated via the BMS on a timed schedule.

The energy efficiency of the building and its operation has been the focus of the building management team. The overall energy savings are tracked on a yearly basis but sub-metering has not been implemented so it is difficult to disaggregate energy use. Comparing the energy bills of the years 2006-07 with those of the previous years, it was possible to estimate an average annual saving of $109,636 equivalent to 1,916,400kWh/year (41kWh/m2.yr). Electrical energy consumption in 2007 was equivalent to approximately 144kWh/m2 as opposed to 185kWh/m2 in 2006. This compares favourably with the US national average electrical energy consumption for office buildings, which is 168kWh/m2.yr[3].

Water consumption is a sensitive issue. Average residential water usage in the USA amounts to over 600litres/person/day (compared with 150 l/p/d in the UK). Typical water consumption in offices for downdraught evaporative cooling is approximately 10 -15 litres/person/day or approximately 1 litre/m2/day. (This actually varies significantly. For example, water consumption for PDEC at the Torrent Research Centre in India is approximately 0.3litres/m2/day). Rainwater is not collected and stored by the building even though theoretically, rainwater collected from the roof (approx. 1,000,000 litres) could satisfy the predicted water requirement of the system (6,000litres/day = 1.2litres/m2/day) for about 6 months, which is roughly the amount of time when PDEC is operational.

Cooling system and building integration

The Passive Downdraught Evaporative Cooling (PDEC) employed in the atrium was included after the schematic design stage, when Arup New York was involved as external consultant in order to overcome predicted risks of overheating in the atrium. The intention was to cool the floor of the atrium and not the whole volume. The balcony corridors were originally part of the transitional spaces, but linear diffusers were added later to thermally separate them from the atrium. Although the designers were aware that stable conditions would not be achieved in the atrium, the threshold set-point for the PDEC system were originally fixed at 23°C for the indoor dry bulb temperature and 60% for the RH. The statistical analysis undertaken by Arup showed that, although for 10% of the time the conditions would have been uncomfortable, that was deemed to be acceptable as the space is transitional. For the permanently occupied spaces in the atrium such as the guards' area, localised cooling was suggested but not implemented.

PDEC PROVIDES A SCREEN OF COOL AIR AGAINST THE OPEN BALCONIES, BEFORE CROSSING THE FLOOR OF THE ATRIUM.

The PDEC system is currently installed on the balustrade of the sixth floor walkway. Although the primary design intent of the engineers was to cool the floor of the atrium[4], a considerable secondary effect is created: a screen of cold air flowing down to the open corridors below and controlling the heat gains from the main atrium space hence reducing the cooling demand of the offices HVAC system[5]. (figs. 84 - 85)

Design-stage predictions suggested an average day-time temperature difference of 11°C would be achieved between the atrium floor and the outside. While this is undoubtedly achieved for much of the summer, significant variation is inevitable. Spot measurements in October 2007 revealed a difference of 7°C between the south end of the atrium and the outside. This difference decreased in proximity of the north glazing where it was only 5°C.

The Phoenix Courthouse's PDEC system is probably one of the largest in the world. It consists of 9 water supply pipes attached to the 6th floor balustrade serving a total of approximately 1100 nozzles. (fig. 87) The nozzles are controlled by motorised valves linked to the building's BMS. The total water flow is 250litre/hour with a pressure of 89bar. At present there are no recorded figures of the daily water consumption but the measurement between the 1st and 2nd of October 2007 show a figure of 6,317litres which makes an average of 263litre/hour (averaged over 24hrs). In a typical hot summer day, water consumptions in excess of 13,000litres have been reported.

During the temperate months when the outdoor temperature provides an acceptable level of comfort, natural ventilation via the east and north openings contributes to removing excess heat gains from the atrium. Fresh air is taken in from the three low level openings above the east entrance, which account for a total opening area of nearly 100m2. The warmer exhaust air is drawn out by stack effect through the 6x30m2 triangular louvered openings at roof level.

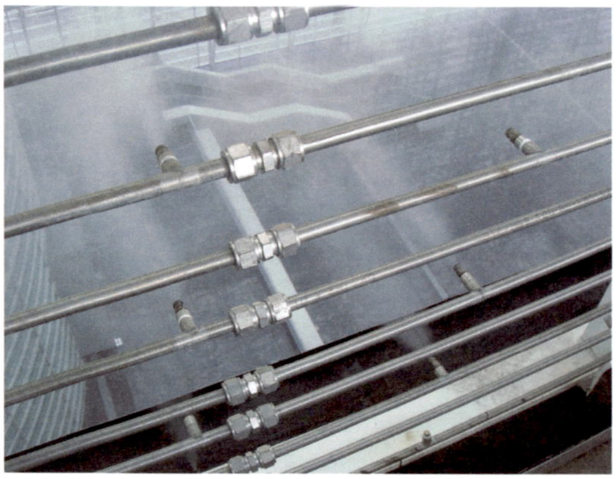

87

84. Main cooling and ventilation strategy Federal Courthouse, Phoenix, Arizona.

85. CFD sectional plot showing temperature stratification in the atrium. Federal Courthouse, Phoenix, Arizona.

86. Atrium.

87. Detail of high pressure nozzles.

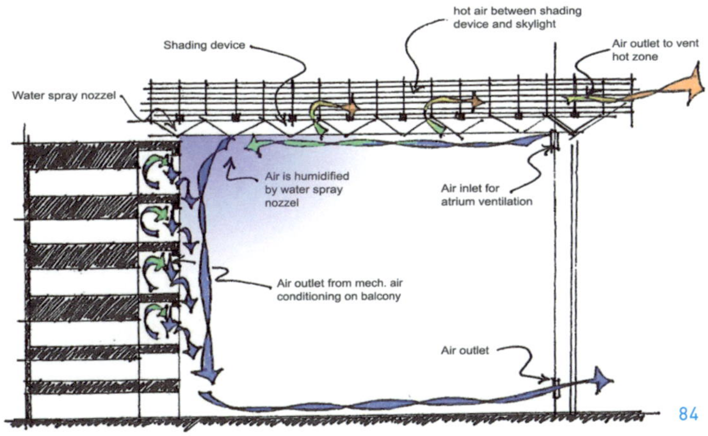

84

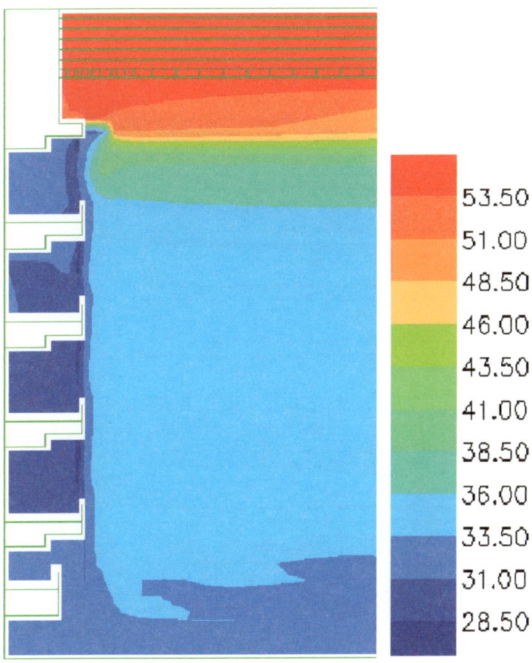

85

86

Occupants' response

The survey involved 19 subjects who answered the questionnaire in the course of one day (2nd October 2007). The summary of results for the post occupancy study is illustrated in figure 88. The subjects were located in the atrium and the surrounding perimeter offices on the ground floor where the impact of the PDEC system is likely to be most significant.

The results show that the majority of the interviewed subjects did not have a very favourable perception of the building, with aspects such a thermal comfort and air quality being the most unsatisfactory. However, care must be taken in interpreting these results, as the sample size is very small, and included many of the security staff members who are permanently located on the floor of the atrium. The respondents have there-

88. Summary of Post Occupancy Evaluation results.

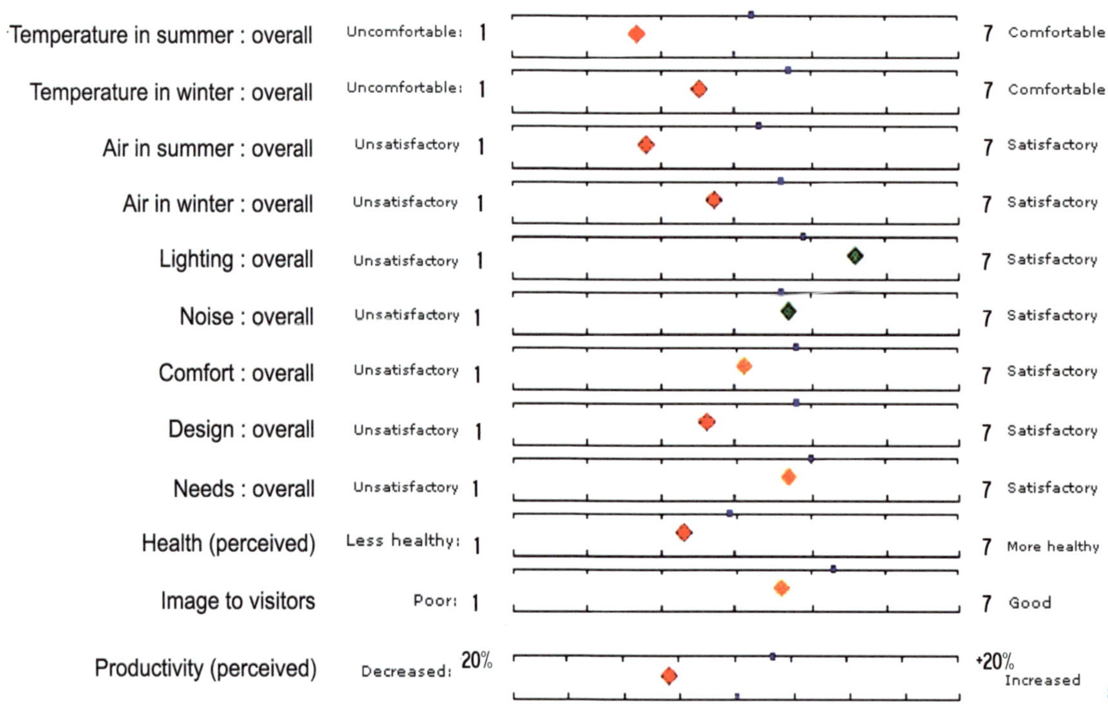

88

fore been reacting to their expectations of a 'working' space rather than a 'transitional' space.

The detailed results on temperature, air quality in summer and control are discussed below.

Temperature in summer: 74% think that it is uncomfortable with 88% perceiving it as too hot and 31% variable during the day.

Air quality in summer: For the majority of the staff air is still, and overall 68% of the sample considers the conditions unsatisfactory. Also, the air was perceived too stuffy by 44% whereas 39% considered it humid. For 83% it was either odourless or neutral.

Control: Between 74% and 79% of the occupants reported they had little or no control on heating, cooling and ventilation confirming the comments made during the informal interview.

The occupants' comments on overall comfort suggest offices and atrium had very different thermal environments: *"It is miserable during the summer outside the office in the atrium of the courthouse. Otherwise it is satisfactory in the office. - It's warm in lobby area but it's cold in FCU building."* Comments also related to the problem of *"Water on floor from misters"*.

The overall building strategy is questioned with several occupants commenting: *"It's a nice design but not for this climate. - It's an amazing design, but not a good design for Arizona's summers. Too much wasted space in the atrium area. - Stylish but no practical"*.

Lessons Learned

Positive Aspects of Project Implementation

PDEC system works well

Following 'fine tuning' of the system, the predicted conditions of temperature and water consumption have now largely been met, and the system works well in tempering the outdoor conditions, providing a cooler environment in the atrium space. The environment on the floor of the atrium provides a successful transition space for most of the buildings users.

Energy Savings

Electrical energy consumption in 2007 was 144kWh/m2. This compares favourably with the US national average electrical energy consumption for office buildings, which is 168kWh/m2.yr[6].

Cost Savings

Passive Downdraught Evaporative Cooling produces a 5M$ saving in the equivalent capital cost of air-conditioning the atrium, and a 75% reduction in annual running costs associated with cooling the atrium.

Problems encountered and opportunities for improvement

The PDEC system experienced problems since the beginning of operation with dripping of nozzles due to blockage and pressure drops.

Following early reports of failure, and although their appointment terminated during design development, the Arup engineers team went back to carry out a post-commissioning survey. Their survey highlighted the following problems[7]:

- None of the 9 rows of pipes above the 6th floor worked on step control as specified.
- The quantity of water coming through the nozzles needed to be reduced.
- Linear diffusers were not working. The air speed of 1m/s

could not be measured and more separation between the atrium and corridor spaces was required. 60cm drop down glass panels were installed near the diffusers (although 1m was originally recommended).

- The control thermostat was wrongly placed in the path of the downdraught.

Originally the 9.5mm diameter water pipes and nozzles were designed to withstand a pressure of 48 to 55bar. By increasing the pressure to 89bar and replacing the blocked Honeywell brass nozzles on a yearly basis, the external contractor, together with the General Service Administration maintenance engineers, report to have successfully addressed the problems of dripping and 'spitting'. This is perhaps surprising considering the relatively high cost of brass nebulisers and the increased water consumption due to higher pressure. At the Torrent Research Centre in India, the resident engineer reported that during the cooling season any blocked nozzles would be removed, cleaned and reinstated on a monthly basis, and this was not regarded as an onerous task.

PASSIVE DOWNDRAUGHT EVAPORATIVE COOLING PRODUCES A 5M$ SAVING IN THE EQUIVALENT CAPITAL COST OF AIR-CONDITIONING THE ATRIUM, AND A 75% REDUCTION IN ANNUAL RUNNING COSTS ASSOCIATED WITH COOLING THE ATRIUM

Another factor contributing to the teething problems of the system was the control logic. This is now regulated so that the nozzles operate only when the external relative humidity is lower than 45%, tightening the earlier control settings. The problem experienced by the occupants of the atrium corridors was dripping and/or condensing moisture on the lower balustrades and the surrounding areas. The perception that the atrium space is 'warm' will be enhanced by the fact that the fully air-conditioned courtrooms and offices are kept at a considerably lower temperature.

High pressure systems require a considerable amount of water and filtration. Currently, the water is purified by a reverse osmosis water filtration system (Nimbus ozone generation system). The filters are changed once a year and the tank is drained every other week in winter. The water is tested once a year but consumption and the system's operation is not automatically recorded by the BMS. The tank capacity is 300 gallons but this is an on-demand system and the water filtering happens by directly injecting the chemical treatment. The old system used twelve 300-gallon tanks storing filtered water but often this was not sufficient. Since the refurbishment of the system, a new drainage system was installed, which has individual drainage pipes, which are more effective when shutting off individual banks of nozzles.

Conclusions

The cooling strategy for the atrium of the Federal Courthouse in Phoenix is a major demonstration of the potential of passive downdraught evaporative cooling which has provided significant savings in both capital and running costs. It is a very large volume, and PDEC on this scale was initially undoubtedly a challenge for the building managers. However, the predicted conditions of temperature and water consumption have now largely been met, and the system works well in tempering the outdoor conditions, providing a cooler environment in the atrium space in the proximity of the south side offices.

Water consumption is higher than other buildings (1.2litres/m2/day compared with 0.3litres/m2/day at the TRC building), but this is in line with the high level of water consumption in the USA (over four times the per capita consumption in Europe).

The views of the occupants (reflected in a small sample survey) identify a general dissatisfaction with temperatures in the atrium in the summer. However, the expectations of the occupants, in a city where shopping malls and public buildings are air conditioned, are likely to have a significant influence on their response to more variable conditions. Not surprisingly perhaps, the rather negative perception of summer thermal comfort was particularly prevalent among the group (security guards) who work for long shifts in the middle of the atrium, with little opportunities for adaptation. It is possible that the inclusion of the security staff in the survey has skewed the results, and many of the staff working in offices may find the atrium satisfactory as a transition space. On the other hand, most of the buildings occupants may have expectations of a fully conditioned space, and thus feel 'dissatisfied' when they experience a wider range of conditionings inside a modern building. This is particularly the case when the home environment, the journey to work and all public buildings are air-conditioned. Further surveys of occupant perceptions are required to obtain a better understanding of 'expectations' during the hot summers of south-western USA.

THE COOLING STRATEGY FOR THE ATRIUM OF THE FEDERAL COURTHOUSE IN PHOENIX IS A MAJOR DEMONSTRATION OF THE POTENTIAL OF PASSIVE DOWNDRAUGHT EVAPORATIVE COOLING WHICH HAS PROVIDED SIGNIFICANT SAVINGS IN BOTH CAPITAL AND RUNNING COSTS

89. View of Federal Courthouse, Phoenix, Arizona.

89

3.3 TORRENT RESEARCH BUILDING
Local Context

The Torrent Research Centre is located on the edge of Ahmedabad within a large 'science park' setting beside the Sabarmati river. (fig. 90) Ahmedabad is the major commercial and industrial city in Gujarat, North West India, and has historically been associated with the production of textiles. Le Corbusier's 'Mill Owners' building is located beside the Sabarmati river a few miles away.

THE LOCAL CLIMATE IS CHARACTERISED BY EXTREMELY HOT DRY SUMMERS (MARCH TO MID-JUNE) WHEN PEAK TEMPERATURES CAN REGULARLY REACH 45°C.

The rapidly expanding modern city continues to be a vibrant commercial centre with a global reach. International corporations have tended to locate at the periphery of the city, avoiding the congestion and pollution of the centre. In the early 1990s, and even more now, energy was a major issue, as demand outstripped supply, resulting in frequent power cuts, and the need for major businesses to own their own back-up diesel powered electricity generators. Although this situation is changing, the need to improve energy efficiency of buildings is now more widely appreciated.

The local climate is characterised by extremely hot dry summers (March to mid-June) when peak temperatures can regularly reach 45°C. (figs. 91 - 92) The hot summer is followed by a warm (32°C peak) and humid monsoon (late June to September), brought by the prevailing south-west winds. To cope with these conditions, conventional air-conditioning has become the norm, although this places a huge burden of both capital and running costs on business. Alternatives to conventional air-conditioning therefore present a major challenge and opportunity.

90. View of the Torrent Research Centre.

91. Monthly diurnal averages for Amhedabad, plotted with Climate Consultant 4.

92. Psychrometric chart for Ahmedabad, plotted with Climate Consultant 4.

90

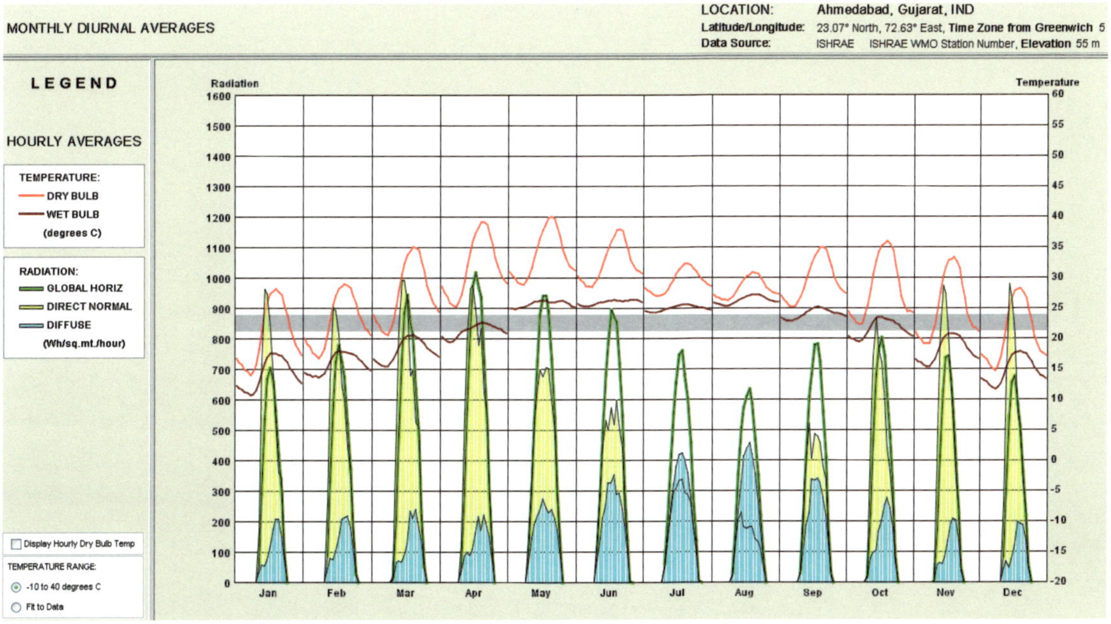
MONTHLY DIURNAL AVERAGES
LOCATION: Ahmedabad, Gujarat, IND
Latitude/Longitude: 23.07° North, 72.63° East, Time Zone from Greenwich 5
Data Source: ISHRAE ISHRAE WMO Station Number, Elevation 55 m
LEGEND
HOURLY AVERAGES
TEMPERATURE:
DRY BULB
WET BULB
(degrees C)
RADIATION:
GLOBAL HORIZ
DIRECT NORMAL
DIFFUSE
(Wh/sq.mt./hour)
Display Hourly Dry Bulb Temp
TEMPERATURE RANGE:
-10 to 40 degrees C
Fit to Data
Radiation
Temperature
Jan
Feb
Mar
Apr
May
Jun
Jul
Aug
Sep
Oct
Nov
Dec
91

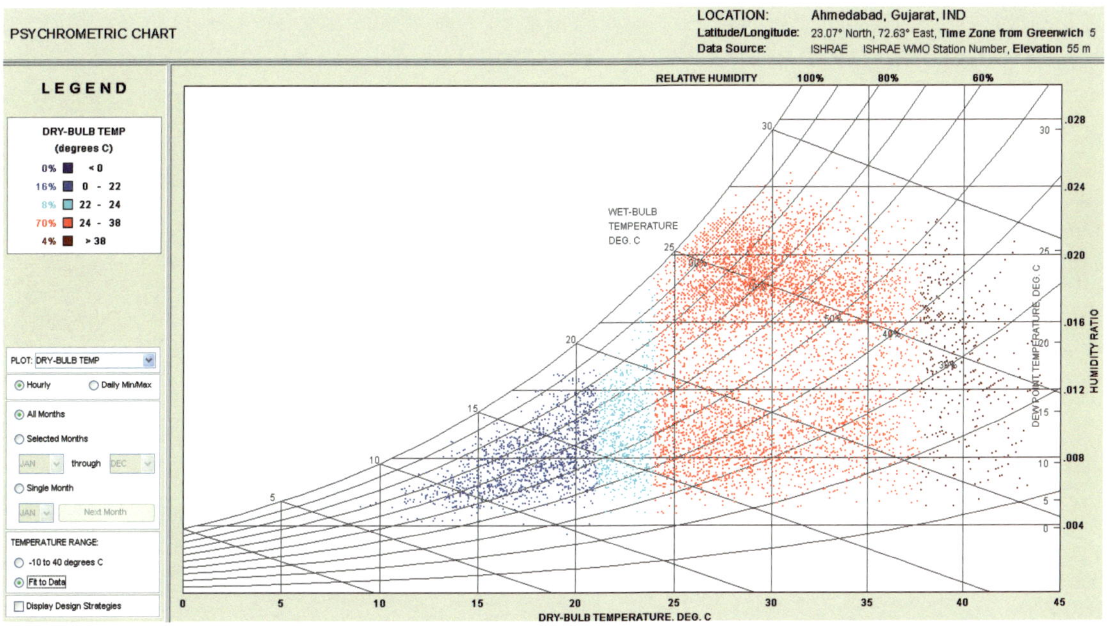
PSYCHROMETRIC CHART
LOCATION: Ahmedabad, Gujarat, IND
Latitude/Longitude: 23.07° North, 72.63° East, Time Zone from Greenwich 5
Data Source: ISHRAE ISHRAE WMO Station Number, Elevation 55 m
LEGEND
DRY-BULB TEMP
(degrees C)
0% < 0
16% 0 - 22
8% 22 - 24
70% 24 - 38
4% > 38
PLOT: DRY-BULB TEMP
Hourly
Daily Min/Max
All Months
Selected Months
JAN through DEC
Single Month
JAN Next Month
TEMPERATURE RANGE:
-10 to 40 degrees C
Fit to Data
Display Design Strategies
RELATIVE HUMIDITY 100% 80% 60%
WET-BULB
TEMPERATURE
DEG. C
DEW POINT TEMPERATURE, DEG. C
HUMIDITY RATIO
DRY-BULB TEMPERATURE, DEG. C
92

The building and the overall strategy

General Strategy

The first phase of the Torrent Research Centre provides six linked laboratory buildings on three floors connected to a central administrative complex, designed to connect with a future second phase. The first buildings were completed and occupied in 1998, while the whole of the first phase was completed in 2000. This project was the first large scale application of PDEC in India. It sought to demonstrate that this approach to cooling is applicable to large complex non-domestic buildings[8].

The brief required over 22,000m2 of pharmaceutical research laboratories and offices, including "unclean" areas like chemical synthesis laboratories, to "very clean" areas like tissue culture, molecular biology and drug design areas. Conventional wisdom would suggest that over 50% of such a new laboratory building would require refrigerant based air conditioning, in order to meet the environmental requirements within the laboratories. However, the client, aware of the compelling benefits of achieving improved energy efficiency in this project, sought to minimise dependence on conventional air conditioning without compromising occupant comfort. In the final complex approximately 70% of the laboratories plus all adjacent administrative areas are cooled with PDEC.

THIS PROJECT WAS THE FIRST LARGE SCALE APPLICATION OF PDEC IN INDIA

Part of the energy strategy was to maximise the use of daylighting, while avoiding the risks of overheating through solar gain. In addition the architects (Abhikram of Ahmedabad) used locally available materials wherever possible. The external masonry walls and roof incorporate the natural mineral 'vermiculite' to reduce thermal conduction, although by European standards the U-value is not very low. Half round ceramic pipes are used in the outer face of the inlet and exhaust shafts to reduce entry of the larger dust particles by creating local turbu-

93. Image of construction crew outside one of the towers TRC, Ahmedabad, India.

94. Interior of misting tower showing micronisers and 'khus' matting, TRC, Ahmedabad, India.

PART OF THE ENERGY STRATEGY WAS TO MAXIMISE THE USE OF DAYLIGHTING, WHILE AVOIDING THE RISKS OF OVERHEATING THROUGH SOLAR GAIN

lence. Motorised louvers are located at roof level within the main shafts to seal the building in the event of a dust storm. The high pressure water supply system (now run at approximately 40 bar) & installation of misting nozzles, was designed by the resident engineer Ramesh Borad, who has subsequently also been responsible for operation and maintenance. (fig. 94)

Environmental Strategy

Typically, laboratories and offices are arranged on three levels either side of an open concourse, which allows the circulation of people between spaces. (fig. 95) This arrangement allows evaporatively cooled air to be introduced (via three 4m x 4m towers) to the occupied spaces at each level, and exhausted via perimeter stacks. The perimeter stacks are designed to exhaust air at both high and low level, depending on the existence (or absence) of wind. In this way, the working areas are both physically and thermally buffered from the external environment. The building envelope has also been designed to minimise solar heat gain while allowing diffuse and reflected daylight penetration.

Summer strategy: During the day from late February to mid-June the building is naturally ventilated (by stack or wind pressure differences) until the outside air temperature is several degrees above internal, when the misting nozzles are activated to induce a downdraught of cool air from the central towers into the concourse. (fig. 96) The cool air drops either side of the circulation 'bridges' in the concourse, and is drawn via 'hopper' vents in the concourse wall, (fig. 98) through the laboratory spaces, to louvred vents in the stacks. At night the airflow path is reversed to flush out any residual daytime heat gains. (fig. 96)

Monsoon strategy: When the monsoon breaks in mid-June, the external temperature drops significantly, but

93

94

95. Plan of typical laboratory block. TRC, Ahmedabad, India.

96. Cross section of TRC showing summer daytime cooling strategy.

97. Cross section of TRC showing night time convective cooling.

98. Laboratory concourse with hopper vents to scoop cool air into the offices and labs.

the humidity rises. In these conditions, the client was happy initially to rely on air movement induced by natural ventilation alone. However, after the first year, following staff comments of 'stuffiness', it was decided to augment natural ventilation with the use of ceiling mounted low-speed fans.

Winter strategy: During the day, natural ventilation (from low to high level) is able to remove internal heat gains, while at night vents at the top of the building are closed to prevent excessive heat loss.

The definition of acceptable threshold conditions was the subject of extensive discussions between the consultants and the client, led by Dr C. Dutt, Torrent's Director of Research. Having discussed the reportedly more flexible attitude of the occupants of 'free-running' buildings to variations in temperature and humidity[9], he agreed to the concept of designing for a threshold temperature of 28-28.5°C in still air conditions, which could be exceeded for a certain number of hours, rather than some absolute value. This was in line with an 'adaptive' approach to thermal comfort. The occupants own perceptions of the acceptability of the internal environmental conditions are discussed below.

THE DEFINITION OF ACCEPTABLE THRESHOLD CONDITIONS WAS THE SUBJECT OF EXTENSIVE DISCUSSIONS BETWEEN THE CONSULTANTS AND THE CLIENT

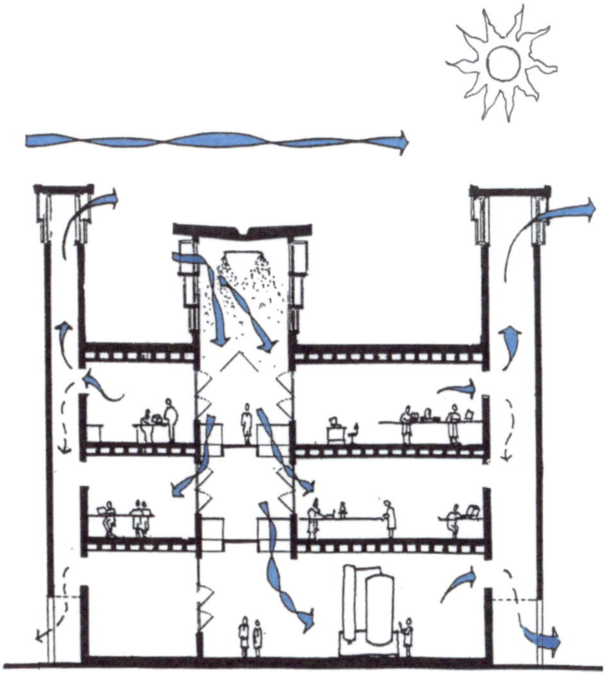

96

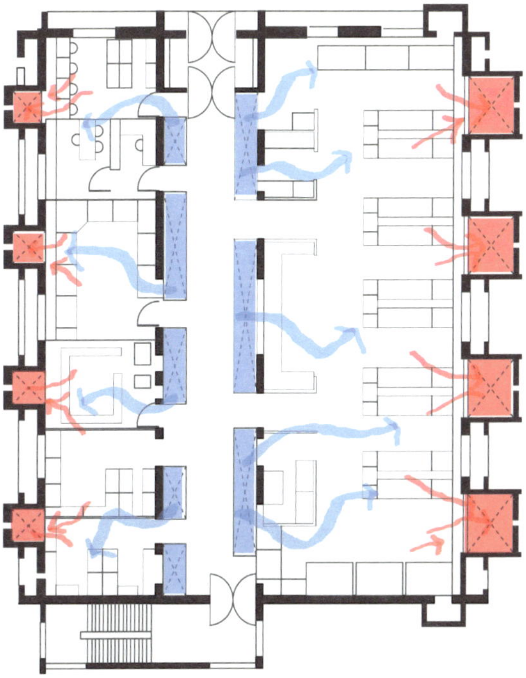

95

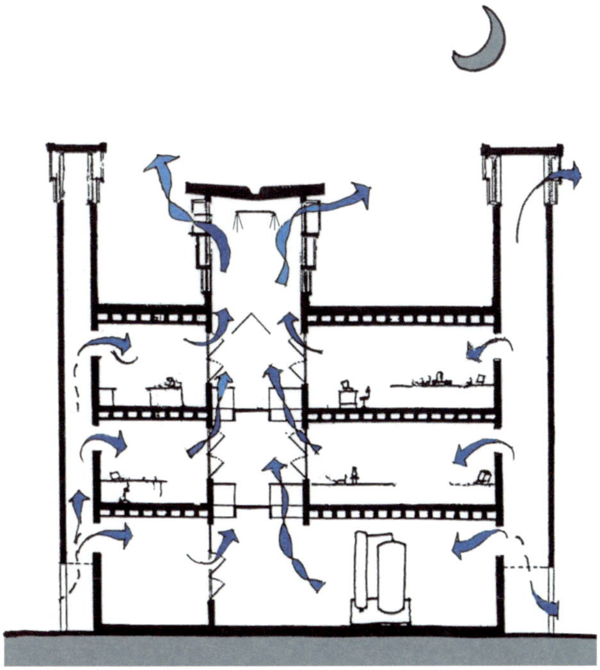

97

98

Cooling System and Building Integration

Each laboratory building has a similar 22m by 17m plan, with a 4m wide corridor flanked by 5m deep office spaces and 8m deep laboratory spaces. (fig. 95) Two of the five laboratory buildings are air conditioned, while the other three are equipped with the PDEC system. The larger main administrative building is located to the north of the laboratories, and a utilities building to the south, with a two level corridor spine linking all the buildings. The entire complex covers 22,600m2 of floor space, of which around 3,200m2 is air-conditioned. The central plant includes two oil fired steam boilers with a capacity of 4T/hr each, two 175cfm air compressors, two 725KVA diesel generator sets, and some 350 Tons of refrigeration capacity.

Overall control of solar heat gains is achieved by careful design of the glazing and by external insulation of the building envelope. The fixed windows (controlled natural ventilation of course occurs via the perimeter shafts) are shaded externally by horizontal overhangs, and in the vertical plane by the air exhaust towers which project from the façade. The buildings are thermally massive - the reinforced concrete construction framed structure has plastered cavity brick infill walls and hollow concrete blocks filling the roof coffers plastered inside. Vermiculite is used as an insulating material on both roof and walls. External surfaces are white - the walls painted, while a china mosaic finish has been applied to the roof.

During the hot dry season mid-afternoon outside temperatures regularly reach above 40°C, while relative humidity is often below 20%. It is under these conditions that the PDEC system is designed to operate. Filtered water is pumped through nozzles at a pressure of 50 Pa to produce a fine mist at the top of the three large (4m x 4m) air intake towers located above the central corridors of each laboratory building. Evaporation of the fine mist serves to cool the air which then descends slowly through the central corridor space via the openings on each side of the walkway. (fig. 96) At each level, sets of hopper windows designed to catch the descending flow, divert some of this cooled air into the adjacent space. Having passed through the space, the air may then exit via high level glass louvred openings which connect directly to the perimeter exhaust air towers. Exhaust air then either rises to the top of the stacks (in windy conditions) or falls out of the bottom of the stack (in still air conditions).

EVAPORATION OF THE FINE MIST SERVES TO COOL THE AIR WHICH THEN DESCENDS SLOWLY THROUGH THE CENTRAL CORRIDOR SPACE VIA THE OPENINGS ON EACH SIDE OF THE WALKWAY

It has been noted[10] that each of the five original laboratory buildings was originally designed for an occupancy of 25 scientists (approximately 15m2/p). With the expansion of activities, increase in staff and overlapping shifts in recent years, some of the buildings accommodate as many as 70 - 80 (approximately 5m2/p) people working at the same time. This will have significantly increased the internal heat gains, and therefore the cooling load. However, the building appears to have met this challenge without a significant rise in complaints from staff.

Measured Performance

A number of studies have reported on the monitoring of thermal performance during the summer period. Monitored temperatures in 1997 and 1998 had indicated that internal maximum temperatures could be maintained 12 to14 degrees below the external peak and that internal temperatures were around 5 degrees lower than average external temperatures[11]. Soon after the first labs were occupied, temperatures were reported of 27°C on the ground floor and 29°C on the first floor with outdoor temperatures at 38°C, and air change rates of 9 per hour on the ground floor and 6 per hour at first floor over the same period. The staff reported that during the summer (Feb - June) the laboratories are comfortable without fans and are not stuffy or smelly, as most chemistry labs are, even when air conditioned. Majumdar[12] reports temperatures of 29-30°C being achieved when outside temperatures reach 43-44°C. Majumdar also reported temperature fluctuations did not exceed a 4 degree range over any 24 hour period, when temperature fluctuations outdoor were as much as 14-17 degrees. One of the early issues noted was a tendency for air to by-pass the top floor[13]. Water consumption for the PDEC system is reported to be approximately 5-6,000 litres/day (approx 0.3l/m2/day).

After the first year of operation the completed building was reported to have used approximately 64% less electrical energy than the equivalent air conditioned building (equivalent to approximately 66kWh/m2.year). Thomas & Baird[14] report that in 2005, overall energy consumption for all four PDEC buildings plus two air-conditioned buildings was 54kWh/m2. This can be compared with the typical energy consumption in Indian commercial buildings which has been reported to be the range of 280-500kWh/m2.[15]. This also compares very favourably to the target for fully air-conditioned office buildings under the recently introduced environmental rating scheme TERIGRIHA (140 kWh/m2.a for day use office building in a "composite" climate). Thomas & Baird[16] point out that "not only is the Torrent building located in a 'hot dry' climate which is more demanding of energy for space conditioning than would be the case in a 'composite' climate, the building has equipment loads which are higher than typical offices and it is used over longer hours than typical office buildings".

AFTER THE FIRST YEAR OF OPERATION THE COMPLETED BUILDING WAS REPORTED TO HAVE USED APPROXIMATELY 64% LESS ELECTRICAL ENERGY THAN THE EQUIVALENT AIR CONDITIONED BUILDING

Occupants' Response

A survey of occupant perceptions was undertaken in 2004 (8 years after the building was opened)[17]. Questionnaires developed by Building Use Studies were circulated to occupants of both PDEC lab buildings and air-conditioned lab buildings, which elicited 164 responses. Most of the respondents had worked in the building for longer than one year.

A total of 292 surveys were distributed and 164 responses returned, 64 from the AC blocks, 100 from the PDEC blocks. Thomas and Baird report a limited difference in terms of the demographics in the AC and PDEC buildings. Respondents in both groups were predominantly male and majority were under the age of 30. Most of the respondents had worked in the building longer than one year, and roughly a quarter to a third of respondents reported that they were seated next to a window. The offices and labs were predominantly open plan with the exception of senior management.

Results of the survey (fig. 99) indicate that both the PDEC buildings and the AC buildings returned mean scores that were *"significantly better or higher than both benchmark and scale mid-point for all of the categories"*. The survey points to *"consistently positive responses with respect to international benchmarks and scale mid-point"* Thomas and Baird[18] also point out that, the co-location of passive downdraft evaporative cooling (PDEC) and air-conditioned (AC) blocks at Torrent offers a unique opportunity to compare performance.

Occupant perceptions of internal temperatures are clearly important in assessing the success of the PDEC system, and the overall results for Temperature Air and Comfort for PDEC in Thomas and Baird's survey corrobo-

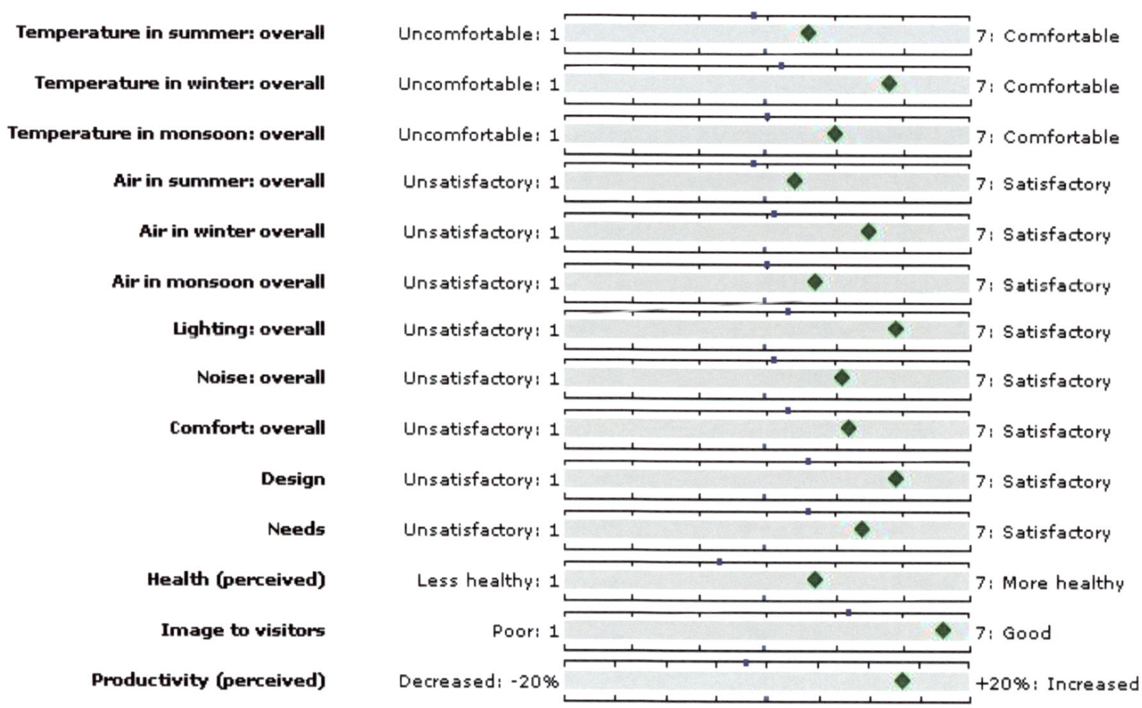

99

99. Summary of Post Occupancy Evaluation results for PDEC buildings at TRC.

rates an earlier report by Majumdar[19] that *"comfort conditions have not been compromised"*.

The summary table indicates that overall occupants have a high level of satisfaction in terms of Comfort, Air Quality and Temperature in the PDEC buildings. In all cases these are above the benchmarks, and compare favourably with the results for the air-conditioned buildings, the conditions of which are also favourably viewed by the occupants. These results for the PDEC buildings are particularly significant in the context of acceptable temperature ranges that are higher than those deemed acceptable in the air-conditioned buildings. These results would tend to confirm the results of field studies undertaken by Nicol et al[20] amongst office workers in Pakistan where office workers were found to be comfortable at temperatures between 20 and 30°C with no cooling apart from fans. (fig. 100)

Although the air-conditioned buildings produced somewhat better results than buildings incorporating the passive downdraft evaporative cooling systems in the BUS survey, it is important to note that the BUS results of the PDEC buildings were also consistently better than international benchmarks and scale mid-points.

Thomas & Baird[21] report that the occupants consistently rated temperatures on the colder side of neutral (mid point in a scale of too hot – too cold) in the air-conditioned buildings. Given that the controlled air conditioned labs are maintained at temperatures around 22-24°C in comparison to generally 5 degrees below the outside mean temperatures in the PDEC buildings, the occupants overall perception is that the PDEC buildings are only marginally less comfortable than the air-conditioned buildings.

The performance of the building in the monsoon season was of particular interest in the PDEC buildings. As

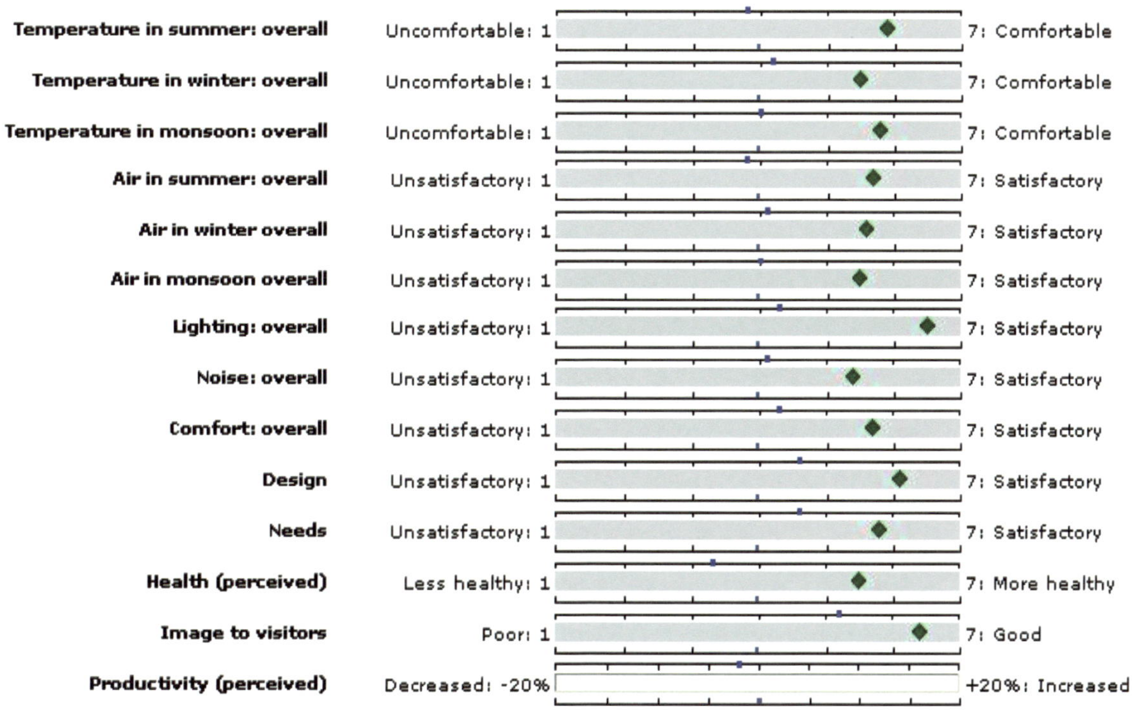

100

100. Summary of Post Occupancy Evaluation results for AC buildings at TRC.

noted earlier, following the first year of occupancy ceiling fans were installed as a consequence of the experience of "muggy conditions" in the building. Occupant responses for this season were generally positive, while occupants experienced some concern about air humidity, and moderate satisfaction with overall air conditions in monsoon. Mean scores for occupant votes regarding acceptable temperatures during the monsoon were close to neutral, indicating their view that, following the installation of the ceiling fans, temperatures were satisfactory.

Open ended comments regarding comfort and ventilation, arising from both PDEC and AC blocks, were predominantly positive. Comments from the PDEC surveys included *"Everything in this building is well equipped for work and comfort"* (PDEC) *"Satisfactory, well ventilated good infrastructure"* and *"Good ventilation"*.

Lessons Learned

Positive Aspects of Project Implementation

PDEC system works well

The passive downdraught evaporative cooling system has been working well since the first building was occupied in 1997. The results of both measured performance and occupant perceptions indicate that the building is providing acceptable conditions at the different times of the year in both the PDEC and conventionally conditioned buildings.

Occupant perceptions very positive

Independent assessments of occupant perceptions indicate that overall, occupants have a high level of satisfaction in terms of Comfort, Air Quality and Temperature in the PDEC buildings. In all cases these are above the benchmarks, and compare favourably with the results for the air-conditioned buildings, the conditions of which are also favourably viewed by the occupants. These results for the PDEC buildings are particularly significant in the context of acceptable temperature ranges that are higher than those deemed acceptable in the air-conditioned buildings.

Substantial energy savings achieved

After the first year of operation the completed building was reported to have used approximately 64% less electrical energy than the equivalent air conditioned building (equivalent to approx 66kWh/m2.year). Thomas & Baird report that in 2005, overall energy consumption for all four PDEC buildings plus two air-conditioned buildings was 54kWh/m2. This can be compared with the typical energy consumption in Indian commercial buildings which has been reported to be in the range of 280-500kWh/m2.a[22]. This also compares very favourably with the target for fully air-conditioned office buildings under the recently introduced environmental rating scheme TERIGRIHA. (140 kWh/m2.a for day use office building in a "composite" climate).

Cost Effective

Although the building cost 4% more than its air-conditioned equivalent, this was paid back in the first year of energy savings. Expenditure on mechanical and electrical plant was reduced by 36%, which has provided further reductions in maintenance costs.

Resident engineer vital to success

The involvement of the resident engineer in the detailed design and maintenance of the pressurised water system has undoubtedly contributed to its success.

Robust performance following increase in occupancy

In spite of the significant increase in the density of occupation of the buildings (from 1 person/15m2 to 1 person/5m2), overall occupant satisfaction with conditions in the building has been maintained.

Problems encountered and opportunities for improvement

Over-spraying

Initial problems of droplets not evaporating fully have been overcome by increasing the water pressure in the system.

Air movement on top floor

Air movement (and volume flow rate) on the top floor of the PDEC buildings is poorer than on lower floors, although this has not been commented on by the occupants. This could be addressed by balancing resistance to airflow on the different levels.

Increase in internal heat gains

The increase in internal heat gains and latent loads from increased occupancy of the PDEC buildings results in more incidences of internal conditions going above acceptable levels, particularly when temperatures peak outside in summer. This emphasises the role of management in understanding the natural limits of the PDEC buildings as originally designed and built.

Conclusions

In drawing conclusions from their survey of occupant perceptions, Thomas and Baird state that: *"the overwhelmingly positive user satisfaction responses of the PDEC blocks coupled with their lower energy consumption validate the integration of alternative climate control systems such as evaporative cooling in contemporary buildings in India."*

In summary:

- The PDEC system works well, delivering satisfactory cooling in summer.
- Occupant perceptions are very positive in comparison with international benchmarks.
- Substantial energy savings have been achieved (total measured electrical energy consumption 54kWh/m2 in 2005, compared with the typical energy consumption in Indian commercial buildings of 280-500kWh/m2.annum).
- Integration of PDEC is Cost Effective (4% additional capital cost paid back in less than five years).
- The resident engineer/building manager has been vital to the successful operation of the PDEC system.
- The buildings and systems have been robust in their performance following an increase in occupant density.

This building, completed over 10 years ago, *"continues to satisfy expectations for a contemporary workplace of high quality that is simultaneously energy efficient."*

INDEPENDENT ASSESSMENTS OF OCCUPANT PERCEPTIONS INDICATE THAT OVERALL, OCCUPANTS HAVE A HIGH LEVEL OF SATISFACTION IN TERMS OF COMFORT, AIR QUALITY AND TEMPERATURE IN THE PDEC BUILDINGS

3.4 MALTA STOCK EXCHANGE
Local Context

The island of Malta (Lat 36°N) lies in the southern Mediterranean, south of Sicily and 290Km from the coast of North Africa. The entire island is comprised of a beautiful honey coloured limestone, which builders have exploited since Neolithic times to construct shelter from the extremes of climate, and to fashion some of the most handsome architecture in Europe. The tradition of high quality masonry construction has continued to the present day.

Malta experiences extremely hot summers (peak temperatures above 40°C are a regular occurrence through July and August), while winters are mild and sunny (rarely dropping below +2°C in January and February). The sea also has a major influence on the climate, and while summer days can be hot and dry, they are more often warm and humid. (figs. 101 - 102)

The site of the Stock Exchange, housed in the former 19th Century garrison church, is very exposed, perched on the south western edge of the bastions of Valletta, and overlooking the Grand Harbour 30 metres below. The elevated position exposes the building to the prevailing north-westerly winds, and also to the (less frequent) south-easterly winds in summer. A characteristic of the north-westerly winds is that they tend to blow during the day in summer, to leave the nights still and calm. The less frequent warm south-easterlies bring dust from the deserts of North Africa, and so ventilation openings must be capable of being closed.

Until recently, like many parts of Europe, the cost of energy for the conditioning of buildings has not been a major issue. However, with the more widespread recognition of the need to reduce carbon and other greenhouse gas emissions, and with Malta's accession to the European Union, buildings will need to comply with EU Directives on energy performance. In the long term it is undoubtedly also in the interests of the Maltese, since (currently) all energy on Malta is imported, and so improved building energy efficiency has become an important goal for everyone on the island.

MALTA EXPERIENCES EXTREMELY HOT SUMMERS (PEAK TEMPERATURES ABOVE 40°C ARE A REGULAR OCCURRENCE THROUGH JULY AND AUGUST), WHILE WINTERS ARE MILD AND SUNNY (RARELY DROPPING BELOW +2°C IN JANUARY AND FEBRUARY)

101

102

101. View of the Stock Exchange, Valletta, Malta.

102. View of the Stock Exchange, Valletta, Malta.

103. Monthly diurnal averages for Malta, plotted with Climate Consultant 4.

104. Psychrometric chart for Malta, plotted with Climate Consultant 4.

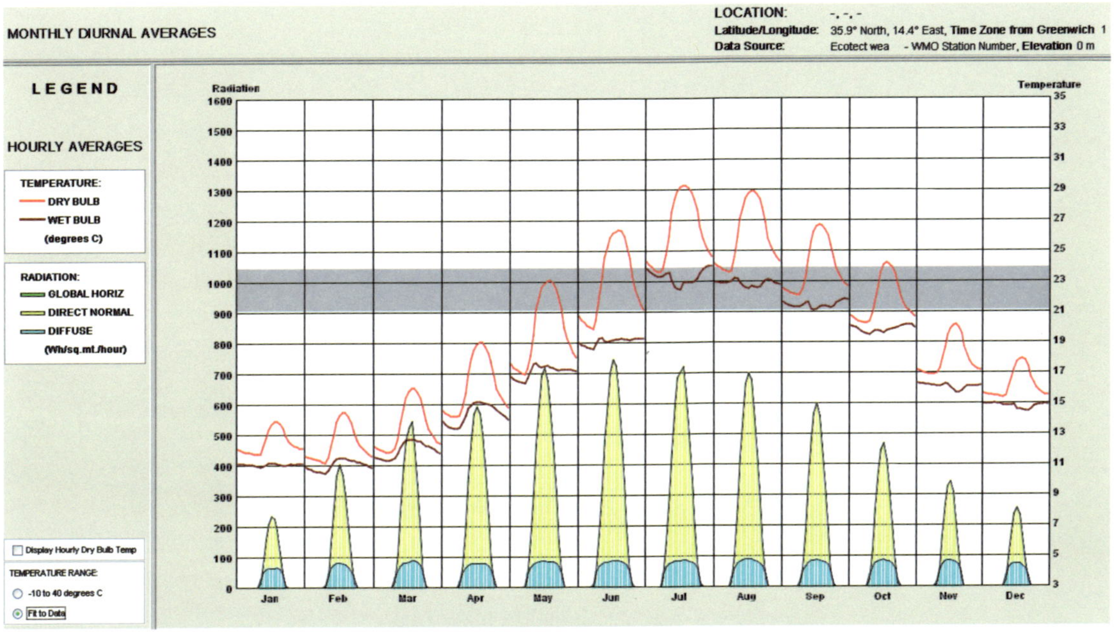

103

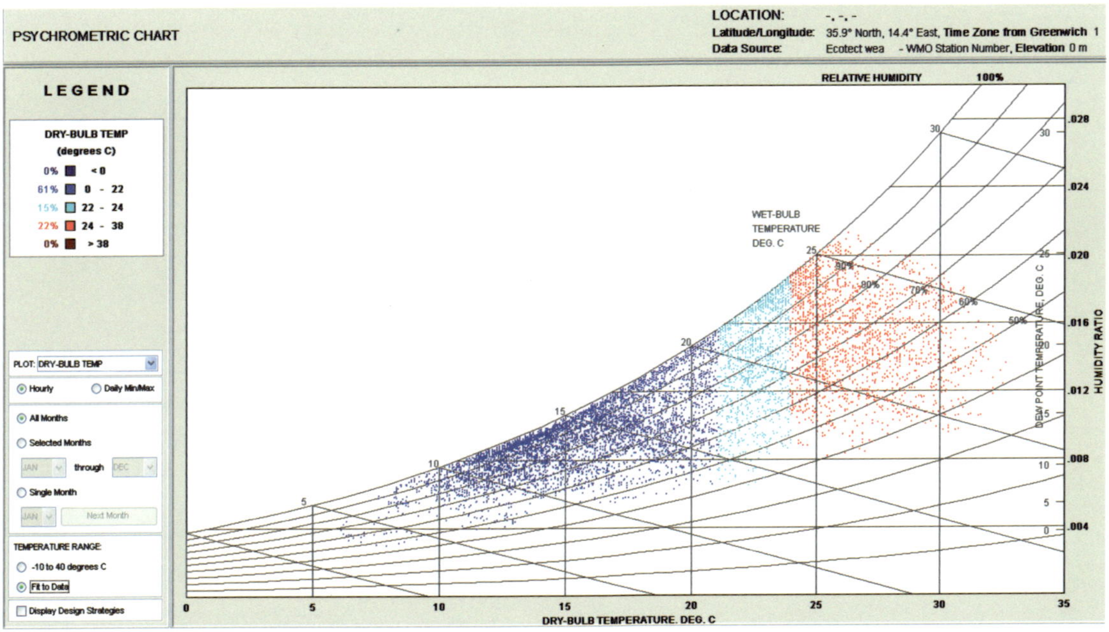

104

The building and the overall strategy

General Strategy

In converting the former garrison church of Valletta into the Malta Stock Exchange, the architects Architecture Project wanted to ensure that the mechanics of cooling (which could have involved extensive ductwork or numerous fan-coil units), would not undermine the architectural qualities of the interior. They therefore sought a solution which would minimise the intrusion of an A/C plant on the interior.

The refurbishment of the existing listed building involved the incorporation of a five-storey atrium surrounded by perimeter cellular offices. This was created partly by digging several metres into the rock below the original floor, and removing the original flat ceiling soffit to reveal the magnificent trusses supporting the roof. The open atrium is also occupied as a work space, creating major challenges for environmental control in the central area of the building. (fig. 105)

Environmental Strategy

The architects were keen to avoid either very intrusive ductwork or large fan coil units within the central 14 metre high atrium. The cooling strategy adopted was therefore designed to minimise reliance on mechanical and electrical equipment through the application of passive and hybrid downdraught cooling. This avoids the visual intrusion of additional plant, and potentially reduces energy and maintenance costs. This same strategy does not apply to the cellular offices and the lower ground-floor meeting rooms, which are cooled with conventional 'cassette' units.

> THE COOLING STRATEGY ADOPTED WAS DESIGNED TO MINIMISE RELIANCE ON MECHANICAL AND ELECTRICAL EQUIPMENT THROUGH THE APPLICATION OF PASSIVE AND HYBRID DOWNDRAUGHT COOLING

During summer days the design proposed to exploit

downdraught cooling, (fig. 107 a) which is induced in a column of air by either evaporative cooling by means of hydraulic nozzles ('micronisers') when ambient conditions are dry, or direct cooling by means of chilled water coils, (fig. 107 b) when conditions are more humid. Previous research had suggested that a combination of these techniques was both technically and economically viable for many locations in southern Europe. The distribution of cool air with both techniques is from high level to low level and the pattern of resulting air temperatures and air velocities within the space is strongly influenced by the geometry of the major spaces.

In addition, during summer nights (when there is little wind) the design strategy was to exploit buoyancy driven night ventilation, (fig. 107 c) whereby air at low night-time temperatures removes heat built up during the day within the walls, floors and roof of the building. The objective in combining these ventilation and cooling strategies was to provide a comfortable internal environment with minimum reliance on M&E equipment. As we shall see below under 'occupant perceptions', a number of factors have meant that this strategy has been only partially successful.

105. Malta Stock Exchange, Plan of upper floor.

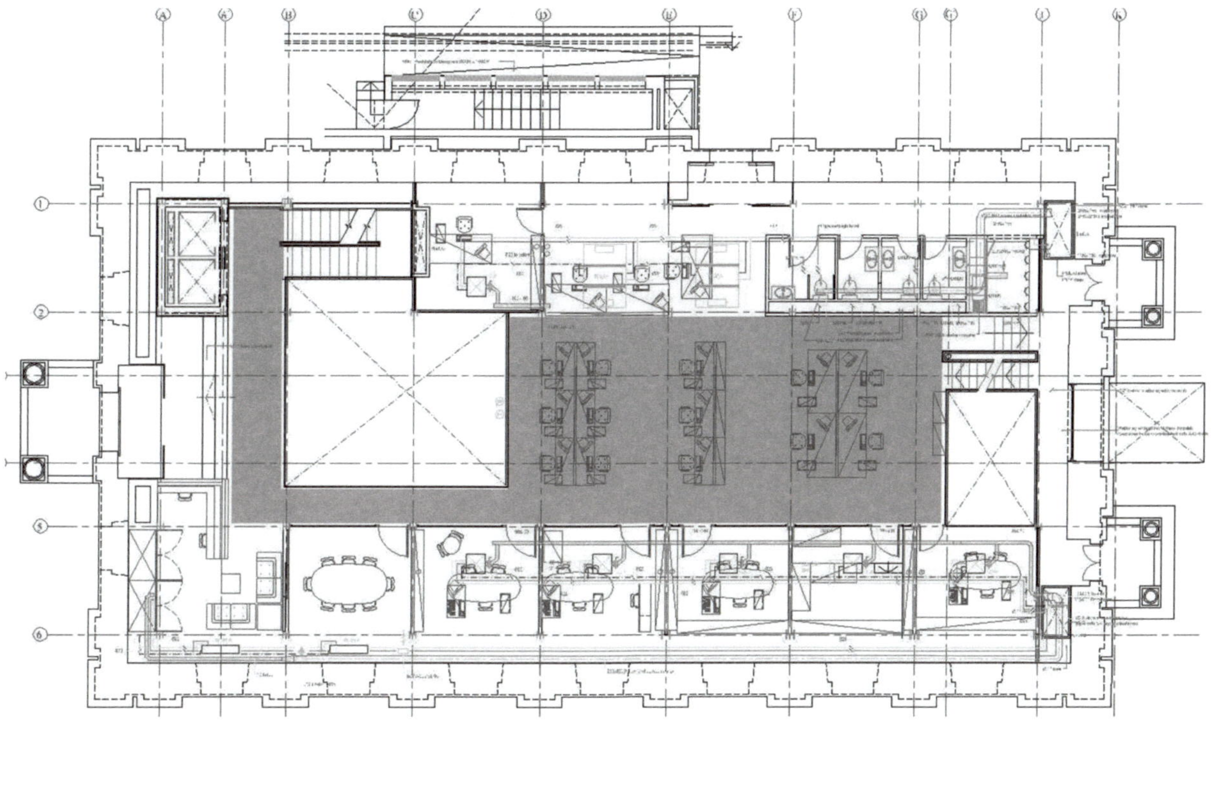

105

Cooling System and Building Integration

The cooling and heating of the large central space was designed to be achieved by five complementary strategies:

1. Automated natural ventilation

The 18 high level vents (1.5x0.85m) are top hung windows at the ridge, with a maximum throw of 0.5m. These window vents are vulnerable to winds from the south west (average 4m/s), so each vent is operated by two steel toothed 'racks' or stays, rotated by pinions located on a tubular drive shaft operated by 3 phase motors. The vents at lower ground level consist of motorised dampers mounted in the east, west and south walls. These ventilators have edge seals to minimise uncontrolled ventilation (although these seals have been found to be ineffective). The operation of the vents is controlled by the building management system, although this can be overridden (and often is) by the building manager (discussed below). (fig. 106)

2. Chiller Based Downdraught Cooling

Two chilled water circuits serve cooling coils from the header along each side of the walkway. The copper coils are inclined at approximately 70° to encourage airflow through the coils, and 300mm wide purpose made trays were installed to collect condensate. The cooling coils call for chilled water in response to variations of both internal temperature and internal relative humidity. The control strategy adopted assumes that the cooling coils are linked to the automatic dampers and vents and operate in conjunction with them. When the coils are switched on, the actuators at high level and the dampers at low level are both set to the minimum aperture.

3. Passive Downdraught Evaporative Cooling (PDEC)

The PDEC system, as designed and installed, relies on 14 hydraulic nozzles at 1pair/bay in four sets. Each set is controlled by motorised valves linked back to the Building Management System (BMS). The total water volume flow required is approximately 90litres/hour with a pressure of 25 bar. The pipe work is attached to the balustrade and the micronisers have been located at the centre of the gap between the cooling coils. The PDEC system also operates in conjunction with the motorised dampers. When the system is on, both banks of motorised dampers and high level vents are set to fully open. The aperture of openings may be overridden by the wind speed/direction sensor. The dampers are set to the minimum aperture when the wind speed is above 5m/s.

Both evaporative cooling and the chilled water coils (figs. 107 a - 107 b) rely on buoyancy to drive the airflow through the building, and so fan power (which can sometimes be 30-40% of the electrical energy used in conventional air conditioning) will be avoided, providing a potentially significant electrical energy saving.

4. Night Time Convective Cooling

It was anticipated that during the day in summer, internal temperatures would rise to 25°C or 26°C. Useful convective cooling is therefore promoted when the external temperature drops below 23°C, which occurs frequently at night during the summer period. The movement of air driven by either wind or buoyancy forces, (fig. 107 c) reverses the daytime air movement pattern under downdraught cooling. Night cooling is controlled via the motorised dampers at lower ground level and the vent actuators at ridge level. Convective cooling relies on large air change rates. When night ventilation is initiated, all low and high level vents are fully opened. In this way the benefits can be genuinely regarded as 'free' cooling.

5. Heating

Fan-coils have been used with the heat pump chillers for a maximum water temperature of 50°C. Simulations indicated a peak demand of 25-30kW to bring the central space up to 21°C from an ambient low of 8°C on a Monday morning in February. However, once the building is occupied the demand for heat is very low. It was important to make sure that the location of these fan-coils unit avoids the obstruction of the airflow path under natural ventilation.

106a-b

106. (a-b) Moveable shading shutters to south windows of Stock Exchange, Valletta, Malta.

106. (a-b-c) From left to right: summer day cooling strategy using PDEC, Cooling Coil and summer night ventilation strategy.

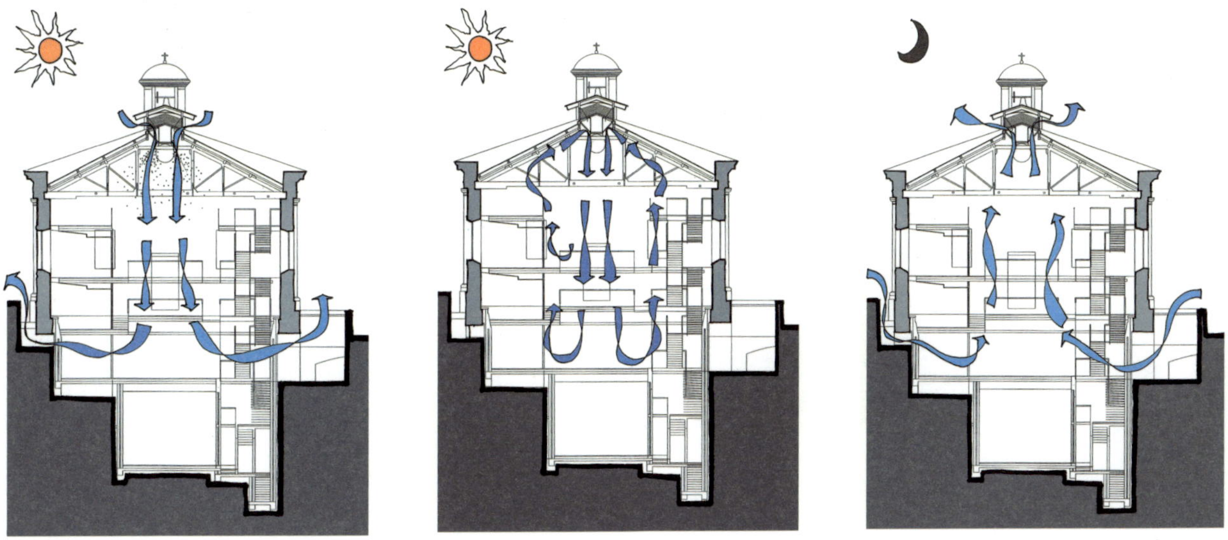

107a-b-c

Design Stage Performance Prediction

Performance prediction at the design stage indicated that temperatures in the main occupied areas of the atrium would remain at an average of 24-25°C during summer daytime occupancy periods, but these vary according to localised internal gains across the plan. Warm air rises from the lower ground floor through the voids on the north side of the building. On the south side the downdraught effect from the cooling coils is more emphasised although with low velocities (between 0.3-0.4m/s). The general result indicated a good mixing of the air across and along the central space and a fairly even temperature distribution. With an outside temperature of 38°C, the downdraught cooling effect from the coils appeared to be able to maintain relatively low internal temperatures with no signs of overheating. (figs. 108 a - b)

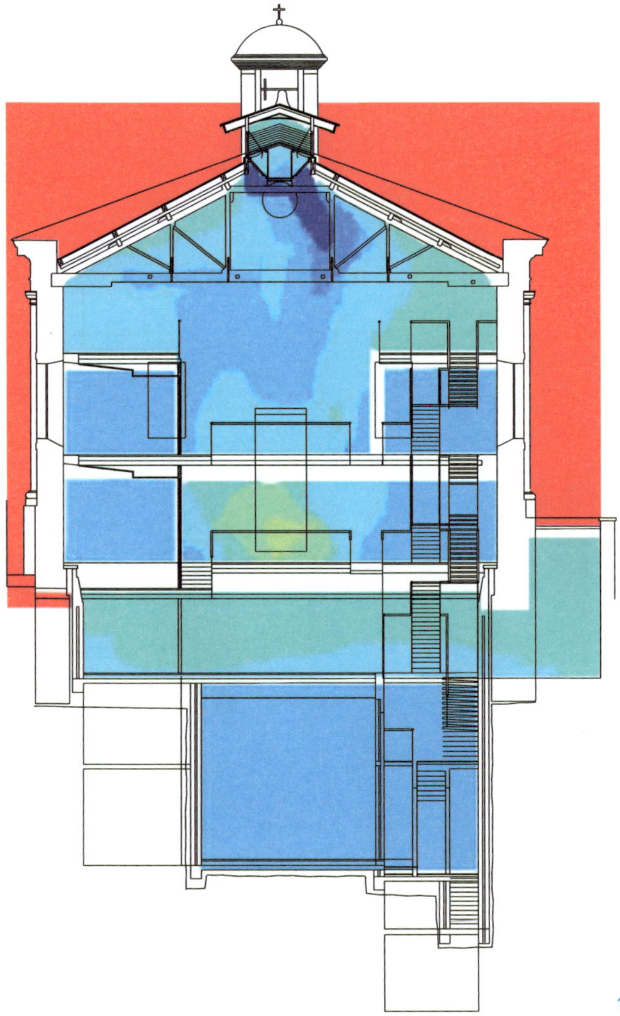

108a

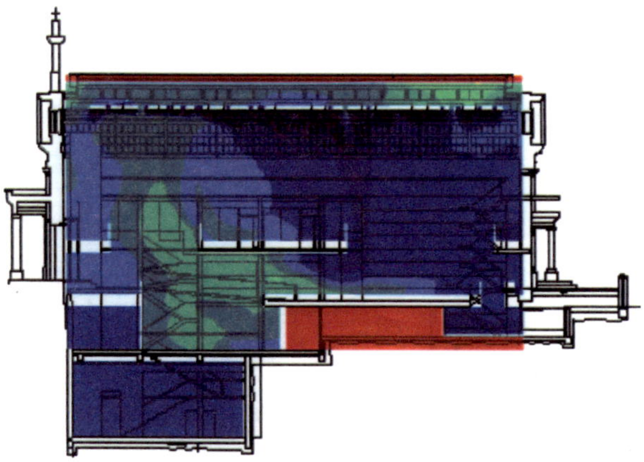

108b

108. (a-b) CFD temperature plot overlay on section of Stock Exchange, Malta.

Commissioning Period Measurements

The building was occupied (in August 2001) before the buildings systems and controls had been fully commissioned. However, a limited set of measurements of the performance of the cooling coil system were made over a period of a few days in September 2001. The results are described in detail elsewhere[23], and concluded that although the cooling coil system was operating in line with predicted performance, during the period of measurements the relative humidity was too high for the evaporative cooling system to be activated. With the high level vents closed, an almost symmetrical pattern of air movement was generated down through the central space, returning up past the balustrades to the roof and through the coils at walkway level. (fig. 109)

The record of continuous measurements shows the impact of the cooling coils on temperatures in the atrium very clearly. Measurements for the 21st September are closest to a 'controlled' test in that there were no 'cooling gains' from the cellular offices and the ground floor door and gable vents were closed. They show the cooling coils bring the temperature at the first floor to just above 24°C at an RH of 70-75%. On the top floor temperatures were 1-2°C higher, which are consistent with theoretical temperature predictions.

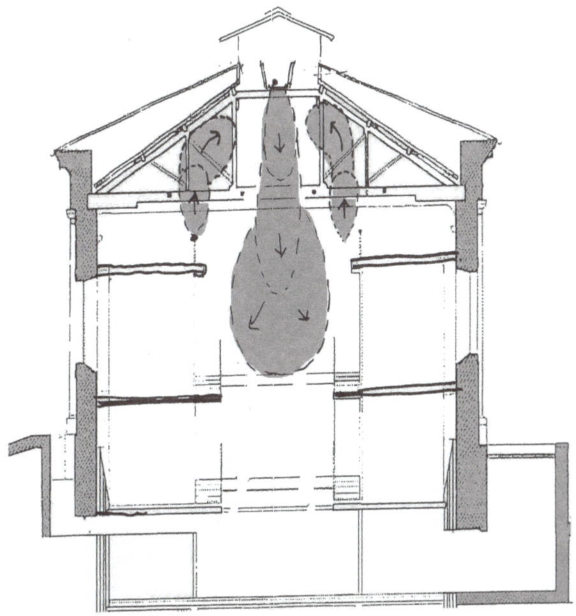

109

110

109. Smoke Test with high level vents closed, showing downdraught from cooling coils and displaced warm air rising to high level before being cooled by the coils.

110. Internal view of the Atrium space with partial view of cooling coils and PDEC system.

Occupants' Response to the building in use

Most of the staff enjoy their working environment. However, the occupants reported problems with the evaporative cooling system from the first year of operation. The problems included droplets of water being discharged beyond the drip tray, falling on work stations and causing considerable annoyance. This is clearly not acceptable, and perhaps understandably the PDEC system was disabled.

However, these problems may be attributed in part to the fact that the building was occupied before the PDEC system could be properly commissioned. It was also discovered that poor alignment of the nozzles with the drip tray, or their close contact with adjacent water pipes and cooling coils had also clearly been a cause of dripping. Some of these problems were identified during commissioning but subsequently not addressed.

Although the PDEC system is not currently in operation, this is not because of an intrinsic technical problem with the system, but because of a poor installation, which may have arisen due to lack of familiarity with high pressure nozzle systems (i.e. the novelty of the system).

Another matter arising from the fact that commissioning took place after the client had moved in, is that the control of the low level vents was not properly regulated, and staff on the lower floors reacted to unwelcome drafts by obstructing the openings. While appearing to solve the problem in winter, the reduced ventilation opening area has resulted in complaints of 'stuffiness' in summer. It is also not clear whether the night ventilation regime is in operation. While the reaction of staff is understandable, full commissioning of the system could have resolved these problems.

A survey of staff perceptions of the internal environment was conducted in May 2008.

Description and results of the survey

A sample of 33 staff, representing 75% of the building's occupants, responded to the survey. Members of staff who responded were located both in the cellular offices and in the atrium open plan space and the majority of them answered the questionnaire in the course of one day.

The results (fig. 111) show that occupants do not have a very positive view of the building, with aspects such as noise and perceived health being the most unsatisfactory. The overall comfort in the building is rated as 'neutral' (or amber in the BUS colour code) as the occupants' response is no lower than scale midpoint but lower than the benchmark. This means that the majority of the occupants gave a mid-scale vote (neither satisfactory nor unsatisfactory). However, this is lower than the benchmark derived from other buildings, where the overall score tends to be placed on the higher end of the scale (i.e. satisfactory). The same trend applied to 6 out of the 12 investigated aspects, with 5 rated as unsatisfactory and only one ('image') satisfactory.

Overall thermal comfort in summer was lower than both benchmark and mid-scale, suggesting that certain areas of the building tend to overheat (indeed just over half the respondents felt too hot in summer). A few years ago supplementary cooling was installed at level +2 of the atrium as this space was to be occupied (previously only used for storage). However, from occupants' comments it was apparent that there is a major problem with air quality and provision of natural ventilation both in summer and winter. Air quality in summer, however, was better perceived (amber) than air quality in winter (red). The main problem in summer

111. Summary of post-occupancy survey results for Stock Exchange, Valletta, Malta.

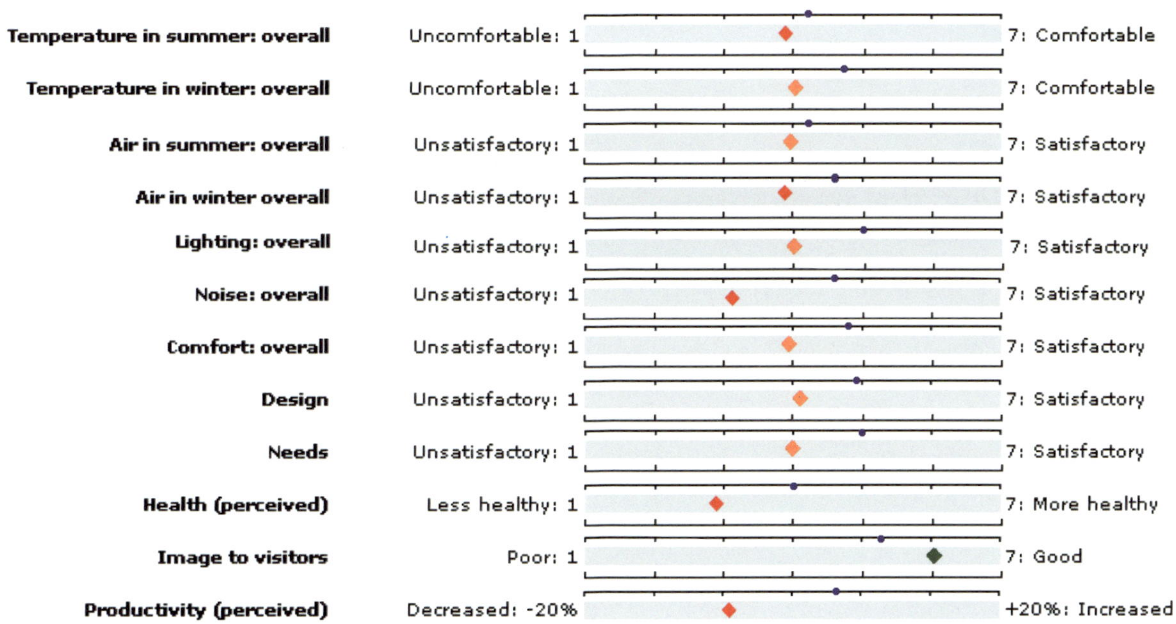

111

seemed to be stuffiness (suggesting inadequate ventilation), whereas in winter it was draught, suggesting over ventilation. It appears that the building manager frequently intervenes to over-ride the control system.

The detailed results on temperature, air quality in summer and control are discussed below:

Temperature in summer: From the survey it appears that 33% think that the temperature is comfortable against 42% perceiving it as uncomfortable. The remaining 25% expressed a 'neutral' vote (i.e. not satisfactory nor unsatisfactory). The overall vote is significantly lower than both scale midpoint and benchmark. However, when prompted on hot/cold perception of the temperature 51% of occupants responded as too hot, which is lower than scale midpoint but not different from the benchmark. Not surprisingly, satisfaction with the temperature in summer depends on where the respondent works within the building.

Air quality in summer: For most of the staff (56%) the air is perceived to be still. It is interesting to note that with the evaporative cooling system disabled, 36% consider the air to be dry, whereas 70% consider it stuffy and 62% smelly (it appears that this latter issue relates to a drainage problem which has now been addressed). This confirms the comments gathered during the informal interviews and the reported problems with the controls of the ventilation system. Overall 45% consider the air quality to be unsatisfactory. This is lower than benchmark but not different from scale midpoint.

Control: Frustration with the environment in the building is apparent, since 80% of the occupants feel they have no control on heating, cooling, ventilation or lighting.

Comments of the occupants: Dissatisfaction with the internal environment is also expressed in comments from staff, particularly with regard to the ventilation: Staff report a *"Lack of fresh air"*, and *"No fresh air"*. The interior is also described as *"Poorly ventilated"*, *"Stuffy air"*. Staff have also complained of: *"Headaches due to stuffiness"*, and that *"More ventilation would be appreciated"*. There are also contradictory comments, some describing the air as *"still"* while others complained of draughts. This is clearly dependent on the location of the respondent in the building. Some employees on lower ground floor, for instance, have obstructed part of the fan coil unit (positioned at high level on their backs) as it was causing draught and discomfort.

Not all the comments are negative. The building is described by staff as: "Modern and elegant", *"Very welcoming and modern design"*, *"Very nice design but in some factors not practical"*.

112

112. Interior view.

Lessons Learned

One of the three elements of the summer cooling strategy for the Malta Stock Exchange has proved to be problematic. Direct evaporative cooling to open plan office areas within the atrium resulted in problems of over spraying and dripping. This was due in large part to misalignment of nozzles (poor installation), Such problems have not arisen at the Torrent Research Centre where direct evaporative cooling via high pressure spray nozzles was also applied, although at Torrent there are no people working directly below the nozzles. At the time the Stock Exchange was occupied, the evaporative cooling system could not be commissioned, and faults with the installation were not discovered until much later.

Night time convective cooling is dependent on automated control of ventilators at high and low level. Motorised dampers at the lower level on the west side of the building have a poor seal and have been a source of unwanted draughts. Unfortunately due to this and other perceived problems, automatic control of cooling and ventilation was disabled.

However, the strategy of using cooling coils at high level has been successful in delivering cooling to the occupied spaces in the atrium as originally designed and forms a precedent for its application in large volume spaces elsewhere.

The premature occupation of new buildings prior to full commissioning of all the mechanical and electrical systems, is a common source of problems (Probe study, 1995-2002). It is also likely that without a dedicated professional building manager, maintenance of the buildings systems and controls has not been properly coordinated.

Energy consumption figures for the building are not available at this time. However, the energy savings predicted at design stage will not have been achieved in full, although the success of buoyancy driven air movement from the cooling coils will have saved fan energy.

The smoke tests revealed a pattern of air movement,

which is also consistent with the CFD analysis undertaken at design stage. Air velocities within the central occupied spaces are low (0.1 to 0.3 m/s), but the movement of large volumes of air at low velocities are desirable and part of the strategy. (Down draught 'plumes' of higher velocity, could become uncomfortable).

However, the combination of detailed air velocity measurement and smoke tests revealed that rising warm air passes over and through the coils where it gets cooled, resulting in a down-draught of cooled air as predicted. Substantial variation in flow over the coils is to be expected, but the overall effect is that the system is dealing with internal heat gains and providing stable, comfortable conditions generally.

Positive Aspects of Project Implementation

Energy saved

Buoyancy driven cooling from the cooling coils will provide an energy saving compared with fan coils or similar air-con units.The general building improvements (shading, insulation, night ventilation) also benefit the cellular office accommodation, reducing the energy consumption of the cassette units by approximately 23% (based on design stage predictions).

The total cooling energy consumption for the whole building using fan coils was estimated at approximately 103 924 kWh (70kWh/m2), compared with 54 139 kWh (36kWh/m2) for the built solution using cooling coils and PDEC, resulting in a total energy saving of 48%. With PDEC not operating, the likely energy savings attributable to the cooling coils in the atrium (i.e. not using fans) will be in the region of 30kWh/m per annum.

Economic benefits

Costs of the original installation. With the system correctly commissioned it should provide maintenance cost savings, but this remains unproven.

Environmental benefits

The estimated energy savings for the building as currently operated represent a reduction in CO_2 emissions of approximately 10,000kgCO_2 (6.5kgCO_2/m2). This is about one third of the savings predicted for the building as originally designed.

Reduced need for bulky plant

The high level cooling coils have been successful in delivering cooling to the atrium space, and this part of the system has worked as predicted. This demonstrates that the downdraught process drives the airflow pattern inside the building, avoiding the need for fans, ductwork and suspended ceilings. Downdraught cooling and night-time convective cooling strategies therefore have clear benefits.

Problems encountered and opportunities for improvement

Energy saved

Energy use for building environment control and energy management has not been recorded, so savings cannot be verified. However, they are estimated to be about one third of those predicted, due to increased dependence on the cooling coils. This is based on the increased electrical load from the compressor.

PDEC system

Although early problems with over-spraying were encountered, this has largely been attributed to poor installation and lack of commissioning. Improvements in the technology of misting nozzles, and experience with buildings elsewhere suggests that an evaporative cooling system could still be viable for the Stock Exchange atrium.

Building Management

Staff have indicated their frustration with the system, and there was nobody who was able to fully comprehend and manipulate the building energy management system. A system of this type will normally require a professional building energy manager to ensure success.

Technical developments have occurred since the system was installed and it is possible that a new misting system could provide trouble free operation.

CHAPTER 4

CAPITAL AND LIFE CYCLE COST

CAPITAL AND LIFE CYCLE COST

The capital and life-cycle cost benefits of adopting PHDC will vary according to location and building type. All components (except for 'Evapcool' porous ceramic panels) are available 'off the shelf', although contractors may add a premium due to lack of familiarity with PDEC systems. In order to assist design teams and clients, Davis Langdon have constructed three life cycle costing models to examine different aspects of PHDC:

- a comparison of options during the design of the Torrent Research Centre in India
- a comparison of various levels of technical solutions from the EVAPCOOL study
- an analysis of bulk data for urban building stock from the Altener study[1].

These models have all been made available for download from the PHDC website http://www.phdc.eu/ (also included in full on the attached CD) and are described in the following sections. The first study is described in more detail in case the reader is unfamiliar with life cycle modelling.

113. Torrent Research Centre, Ahmedabad.

113

4.1 THE TORRENT RESEARCH CENTRE, AHMEDABAD, INDIA

(fig. 113) Prior to the final design of this building a feasibility study was carried out in 1994 to compare three options:

- Option A: a conventional building with air-conditioning
- Option B: a conventional evaporatively cooled building (through cooling pads in the inlets and fan driven ducted supply of air);
- Option C: a sealed evaporatively cooled building with PDEC towers.

DL updated the costs to present day values in £ and set up an LCC analysis that showed that the PDEC system would become the least cost option after only two years. Extracts from the model are shown here. (fig. 114)

Figure 114 shows the main parameters which can be altered to test various scenarios.

By discounting future costs a total cost can be built up over a period to include the initial capital cost and the future running costs. The total cost represents the amount of money that would be needed to be invested now and if drawn upon as required, would pay for all present and future costs over the period under consideration. This total cost is a useful figure for comparing projects that have different future costs occurring at different times. It is assumed that the invested money earns interest

114. Key parameters adjustable in the model.

Torrent Research Centre, Abhikram
Option comparison

Cost update to 2008/9

Key parameters	
Analysis period (for chart)	30 years
Discount rate	3.50% pa
Electricity cost*	6p/kWh
*increase above inflation	0.5% pa

114

and that future costs will increase (or decrease) in line with inflation.

The discount rate is the interest rate divided by the inflation rate. A company will often calculate its own discount rate based on the rate of return on capital rather than using general interest rates. The future cost line (the cumulative cost to date) rises steeply to start with and then curves towards the horizontal; the higher the discount rate the sooner this cost horizon is approached. As the line approaches horizontal, future costs cease to have any significant effect on the overall project cost. This explains why many companies adopt a very short term strategy when considering financial options.

THE LIFE CYCLE COST ANALYSIS FOR TORRENT RESEARCH CENTRE SHOWS THAT THE PDEC SYSTEM BECOMES THE CHEAPEST OPTION AFTER ONLY TWO YEARS.

We have allowed electricity costs to be increased by a set percentage above inflation (inflation is accounted for in the discount rate).

Table 2 explains how the model works. The capital costs, updated from the original 1994 cost estimate, are broken down into the broad categories relevant to this comparison, and set out under year 0. Future maintenance and replacement costs have been calculated as percentages of the capital cost and are entered into the relevant future years; only years 1 and 2 are shown here. All the costs in this table are shown in today's values. Electricity costs are shown rising by 0.5% per year. Electricity costs are shown rising by 0.5% per year.

The total cost in each year is converted to a net present value (npv).

The discount rate in this example has been set at 3.5%, based on inflation of 3.4% and a rate of return of 7%.

The calculation is 1.07/1.034 = 1.035

For example in option a) the total cost in year 2 is £147,557 in today's values. The actual inflated cost will be £157,705.

i.e £147,557 x (1+3.4%)^2

However the discounted cost shown in the table, £137,746, earning 7% per year will accumulate to £157,706 and exactly cover the inflated cost.

Figure 115 shows the total discounted cost for each of the option is over a 30 year period. Building costs are similar although the cooling pad option has the lowest installation cost, while the running costs represent under 50% of the build cost for the PDEC option but nearly 100% with air conditioning.

Figure 116 shows the year by year cumulative discounted cost and this shows that PDEC becomes the cheapest option after only 2 years, and that the cumulative benefits of adopting PDEC are very significant. This updated cost model will be directly relevant to non-domestic buildings in north India.

115 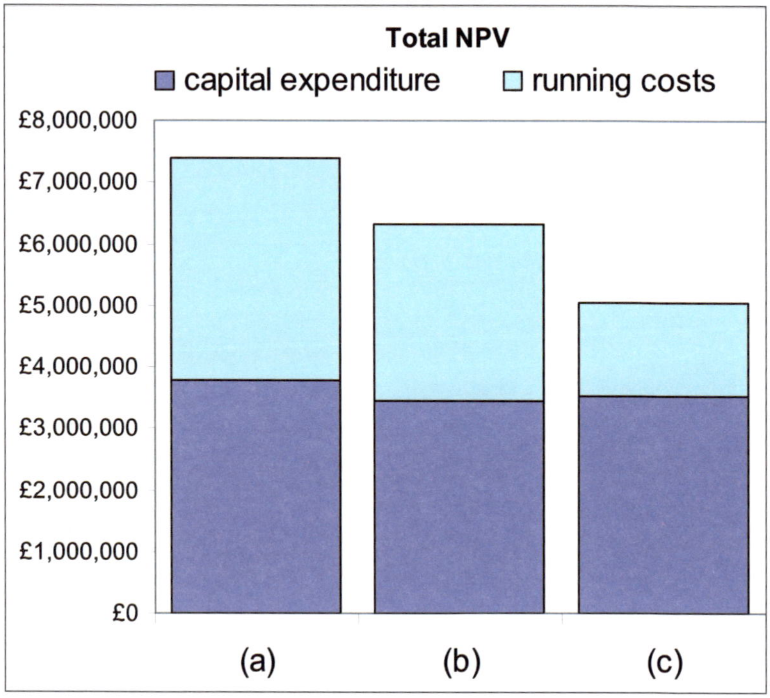

116 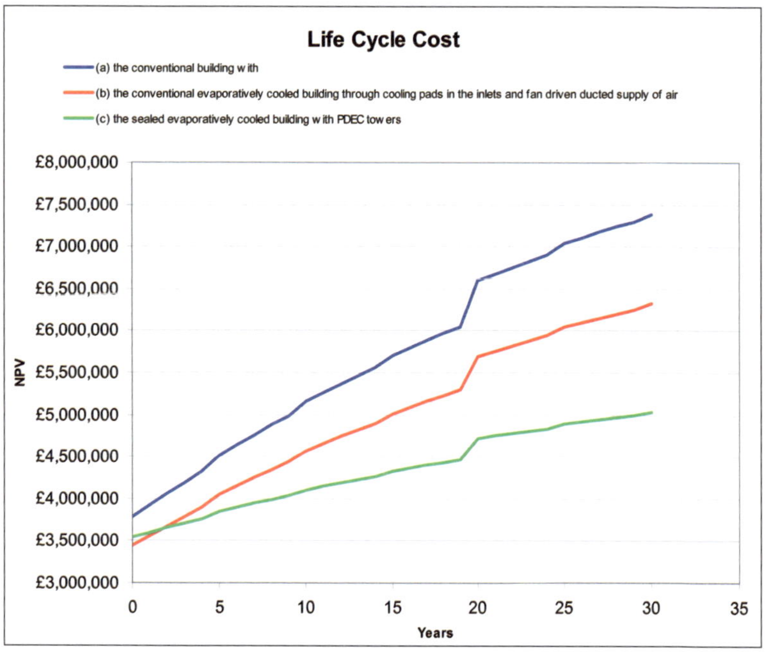

115. Total net present values over 30 years for options a, b and c.

116. TRC Cost Model.

Table 2. Life cycle model for TRC, showing accounted for key variables.

(a) THE CONVENTIONAL BUILDING WITH AIR-CONDITIONED/AIR-EXHAUSTED AND OPEN WINDOW AREAS

		NPV of Future Costs - CAPITAL COST		
	Years	0	1	2
BUILDING COST	Builing works	2,500,340	25,003	25,003
	Insulation	202,370	-	-
	False ceiling	100,710	1,007	1,007
COOLING SYSTEMS	AC plant	839,600	-	-
	Cooling pad			
	Micro-ionisers			
	Fans	16,380	-	-
ENERGY	Electricity	120,340	120,942	121,546
	TOTAL COST	3,779,740	146,952	147,557
	NPV	3,779,740	141,983	137,746
	CUMULATIVE NPV	**3,779,740**	**3,921,723**	**4,059,469**

(b) THE CONVENTIONAL EVAPORATIVELY COOLED BUILDING THROUGH COOLING PADS IN THE INLETS AND FAN

		NPV of Future Costs - CAPITAL COST		
	Years	0	1	2
BUILDING COST	Builing works	2,500,340	25,003	25,003
	Insulation	202,370	-	-
	False ceiling	46,820	468	468
COOLING SYSTEMS	AC plant	390,850	-	-
	Cooling pad	197,700	-	-
	Micro-ionisers			
	Fans	4,960	-	-
ENERGY	Electricity	96,270	96,751	97,235
	TOTAL COST	3,439,310	122,223	147,557
	NPV	3,439,310	118,090	114,548
	CUMULATIVE NPV	**3,439,310**	**3,557,400**	**3,671,948**

(c) THE SEALED EVAPORATIVELY COOLED BUILDING WITH PDEC TOWERS

		NPV of Future Costs - CAPITAL COST		
	Years	0	1	2
BUILDING COST	Builing works	2,817,160	28,172	28,172
	Insulation	241,660	-	-
	False ceiling	42,270	423	423
COOLING SYSTEMS	AC plant	357,760	-	-
	Cooling pad			
	Micro-ionisers	45,050	-	-
	Fans	4,960	-	-
ENERGY	Electricity	30,900	31,055	31,210
	TOTAL COST	3,539,760	59,649	59,804
	NPV	3,539,760	57,632	55,828
	CUMULATIVE NPV	**3,539,760**	**3,597,392**	**3,653,219**

T2

4.2 THE EVAPCOOL STUDY

The EVAPCOOL[2] report, downloadable from the PHDC web site (and available in full on the attached CD) set out to examine the potential use of porous ceramic panels in downdraught cooling systems. Water within porous ceramic containers slowly diffuses to the outer surface where it evaporates in the hot and dry prevalent air stream, loses heat and creates a cool humidified down draught that can be delivered to the indoor space. (fig. 117)

Single ceramic container and stacked wall

The study considered 5 residential options of increasing degrees of sophistication ranging from a very basic low cost installation that might be used in a developing country to an option suitable for a modern home in southern Europe:-

- Vernacular
- Low-cost
- Medium-cost
- High-cost

The LCC analysis compares these installations with the use of a portable air conditioning unit.

As in the last model the key parameters can be changed to produce different scenarios. Here we are using a discount rate of 2% which could represent an earning potential of 5% and inflation of about 3%. Electricity costs have again been assumed to increase at 0.5% per annum above inflation. In this model the country can also be changed and for this example we have selected China and the costs are shown in Renminbi.

The key parameters are shown above the chart in figure 118. This model compares costs for the EVAPCOOL options to a portable air conditioning unit, a 2kW single split unit using eco-refrigerant, and it is necessary to set the number of hours per day and days per year that the unit will be used.

Table 1 shows a breakdown by components showing costs to purchase and install the different options.

Figure 118 provides a breakdown by installation cost, fuel cost and maintenance costs over a 30 year period. The options have been ordered by increasing capital cost and this demonstrates how important it is to look at more than the initial purchase price.

Figure 119 shows the cost profile for each option, starting with the installation cost in year 0 followed by the cumulative cost to each year in the 30 year profile. The steps in the lines are the routine maintenance and component replacements that will be carried out to keep the systems running at maximum efficiency.

We recommend downloading this model and altering the key parameters to test different assumptions.

117. 'Evapcool' basic porous ceramic stackable component.

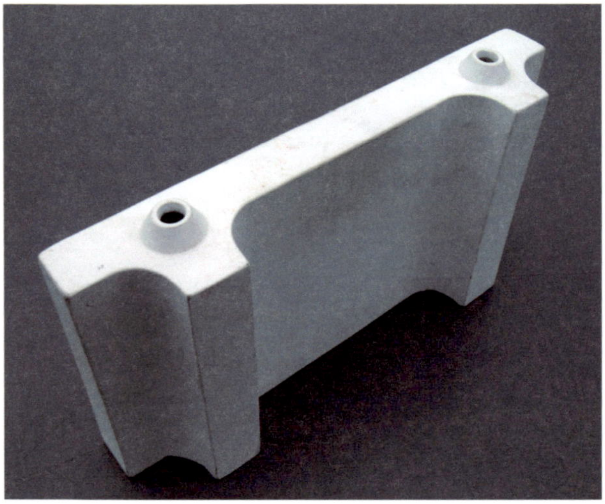

117

118

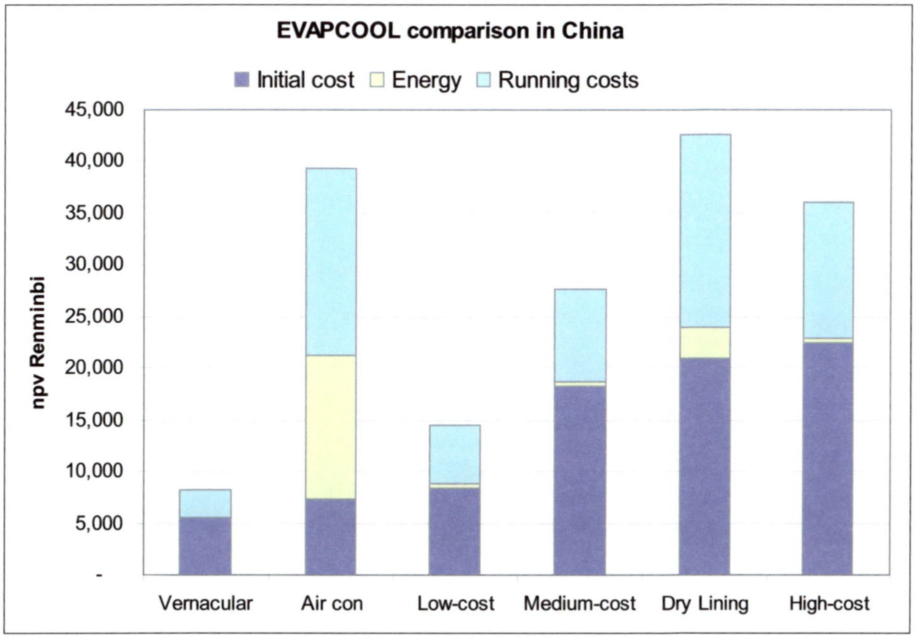

119

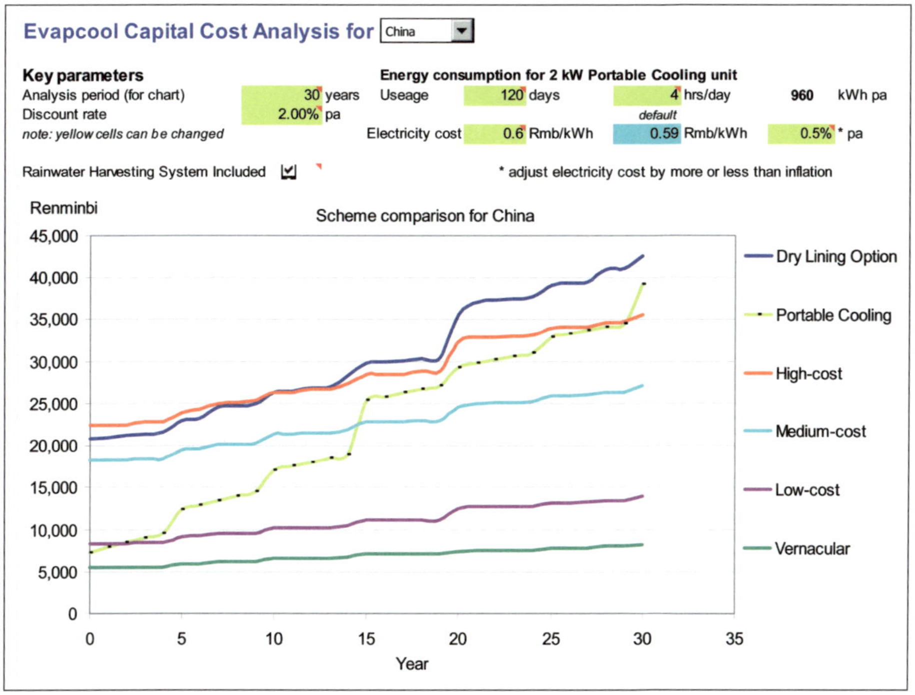

118. Total NPV costs after 30 years.

119. The cost profiles.

4.3 PDEC (MISTING) SYSTEMS

The Altener study included morphology surveys of sample areas from a number of European cities to establish what proportions of existing buildings were likely to be suitable for application of PDEC intervention. Data was collected for more than 2000 buildings and DL set up a cost model to estimate the costs for the various interventions.

The costs in this model have been updated and its scope increased to cover costs from the China, India, Portugal, Italy, Greece, Spain and the UK, to provide a full life cycle estimate and to compare the results to comfort cooling.

KEY LIFE CYCLE PARAMETERS MAY THEN BE ENTERED AND THE MODEL WILL ESTIMATE THE TOTAL CAPITAL COST, THE COST IN EACH YEAR AND THE COST PROFILE.

The model, downloadable from the PHDC website (and available in full on the attached CD) allows basic building dimensions to be entered and determines the appropriate

NPV Costs v Savings

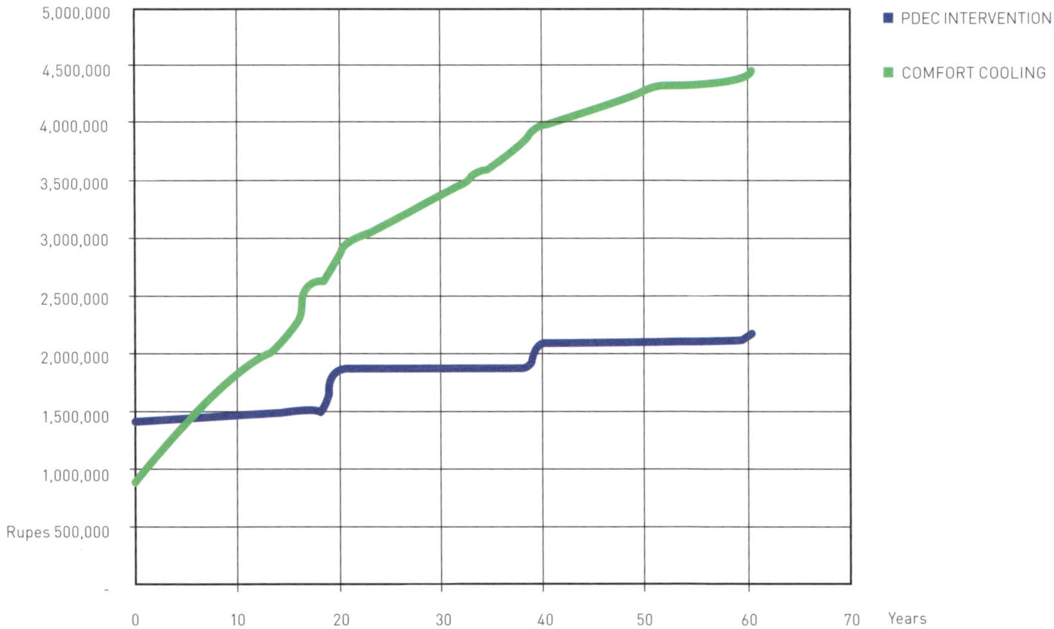

CAPITAL COST - NPV OF FUTURE COSTS - RUPEES

	Years	0	1	2	3	4	5
PDEC	Total NPV (NPV) of cost in year	1,360,065	3,422	3,307	5,806	3,087	23,954
	NPV of cost per m2	2,285.82	5.75	5.56	9.76	5.19	40.26
COMFORT COOLING	Total NPV (NPV) of cost in year	843,115	115,568	111,994	108,532	105,177	101,926
	NPV of cost per m2	1,417.00	194.23	188.23	182.41	176.77	171.30

120

120. PDEC cash flow (Rupees).

PDEC Intervention Costing in India

India ▼

Building parameters

Area per floor	**119**	m2
Floors served	**5**	no
Building depth	**15.00**	m
Average void area	**9.00**	m2
Number of voids	**1**	
Void to elevation	**8.00**	m2
Floor to floor height	**3.5**	m
PDEC rating	1.00	
Tower/void side	3.00	m
Tower/void area	9.00	m2
No. of voids/towers	1	
Additional cooling	☒	
Window actuators	☐	
Discount rate	**3,5%**	pa
Energy saving	28.80	kwhpa/m2
Energy cost*	4.02	Rupees/kwh
Alternative		energy cost
Energy cost increasing	**0,5%**	pa above inflation

T3

Table 3. PDEC Intervention costs in India.

Table 4. PDEC LCC model Inputs and summary.

USER NOTES:

- Insert building dimensions in the blue cells
- Adjust general parameters in the yellow cells
- Select princing location from drop down box
- Options to install window actuators for intermediate
- Intervention and to connect an existing cooling system

Minor intervention

Cumulative NPV over 30 years including capital costs

PDEC Intervention		
	Total cost	1,856,562 Rupees
	Total cost per m2	3,120.27 Rupees
Comfort Cooling		
	Total cost	3,422,253 Rupees
	Total cost per m2	5,751.69 Rupees

T4

intervention (if any). Key life cycle parameters may then be entered and the model will estimate the total capital cost, the cost in each year and the cost profile. (Table 3)

The following example is for a small 5 storey city centre office in India with costs given in Rupees.

(Table 4 - fig. 120) show details of the model inputs and outputs. The chart demonstrates that for the chosen parameters PDEC will become the cheaper option after about 5 years and will cost half as much after 40 years. For this particular building the model has established that a Minor Intervention will be appropriate. The model also shows a more detailed costing for each intervention and separately for the spray towers.

By running all the data collected from the morphology survey through the LCC model a more comprehensive picture of applicability was built up, see figure 10. In total 2,393 buildings were surveyed and of these just over 80% were suitable for one of the three intervention levels:-

- Minor - constructing a PDEC tower over an existing lightwell.
- Intermediate - attaching a PDEC tower and shaft to the side of a building.
- Major - breaking through floors to create a new lightwell and construction of PDEC tower. This option is also suitable for new build.

The key parameters were set to:

- 3.5% discount rate
- energy cost 0.10 € / kwh
- energy cost increases in line with inflation

Minor interventions have the lowest installation cost and are suitable for about 7% of buildings. 50% of them achieve payback, i.e. start saving money compared to comfort cooling, within 2 years and 70% within 13 years.

Intermediate interventions are suitable for 32% of buildings and 40% of these achieve payback within 12 years and 60% within 40 years.

Major interventions are the most expensive and are the only option in 42% of buildings. 30% of these achieve payback within 14 years but 60% are within 33 years and this is better than intermediate interventions.

Given long enough, around 70% of interventions will prove cheaper than comfort cooling. (fig. 121)

Figure 122 shows the same data but with fuel costs increasing by 5% per annum more than inflation. This brings forward the payback periods and 50% of all interventions now achieve payback within 15 years instead of 25 years and 80% instead of 7% will eventually achieve this target. Fuel costs are likely to rise faster than inflation but by some amount between these two examples.

Conclusion.

Life-cycle costing models for PDEC systems have been developed by David Langdon Consultants to enable analyses to be undertaken for different locations and different market conditions. These models are available from the PHDC website (http://www.phdc.eu/) and on the attached CD. The general situation (April 2010) suggests that PHDC systems are highly competitive against conventional systems in India (achieving a simple payback with 2 -5 years), and are also competitive in Southern Europe and China. Assessments of the applicability of PHDC to existing building stock in these three areas, and potential barriers to the development of a market in PHDC systems follow.

121. Running the morphology survey data through the LCC model.

122. increasing fuel costs by more than inflation.

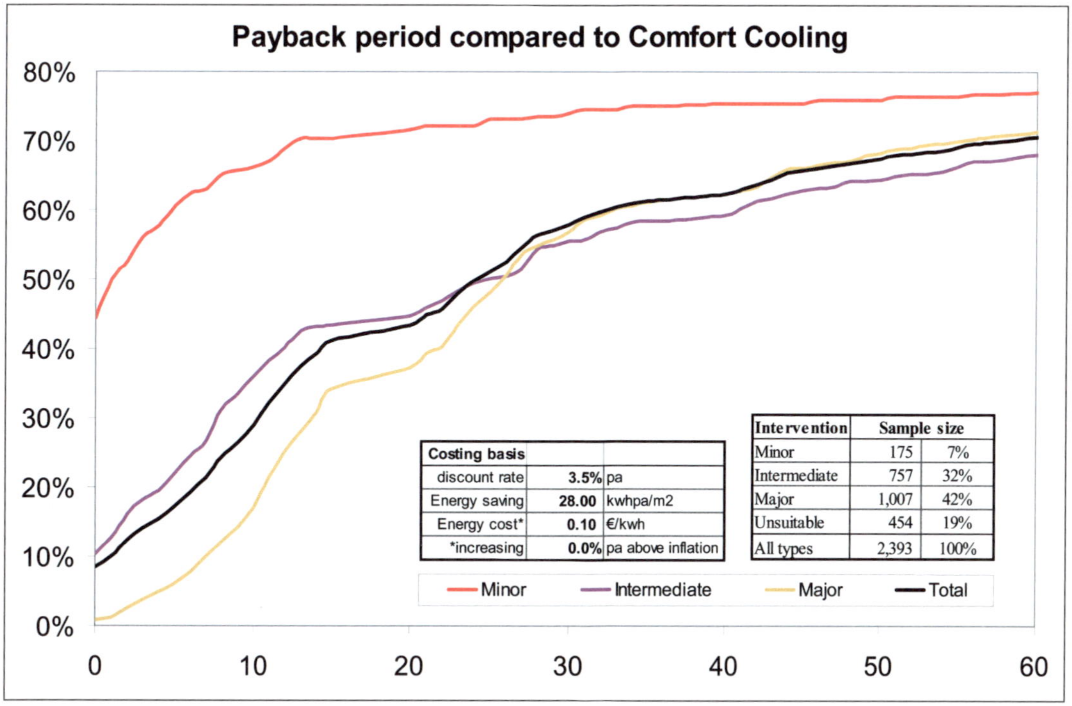
Payback period compared to Comfort Cooling
80%
70%
60%
50%
40%
30%
20%
10%
0%
0
10
20
30
40
50
60
Costing basis
discount rate 3.5% pa
Energy saving 28.00 kwhpa/m2
Energy cost* 0.10 €/kwh
*increasing 0.0% pa above inflation
Intervention Sample size
Minor 175 7%
Intermediate 757 32%
Major 1,007 42%
Unsuitable 454 19%
All types 2,393 100%
Minor
Intermediate
Major
Total

121

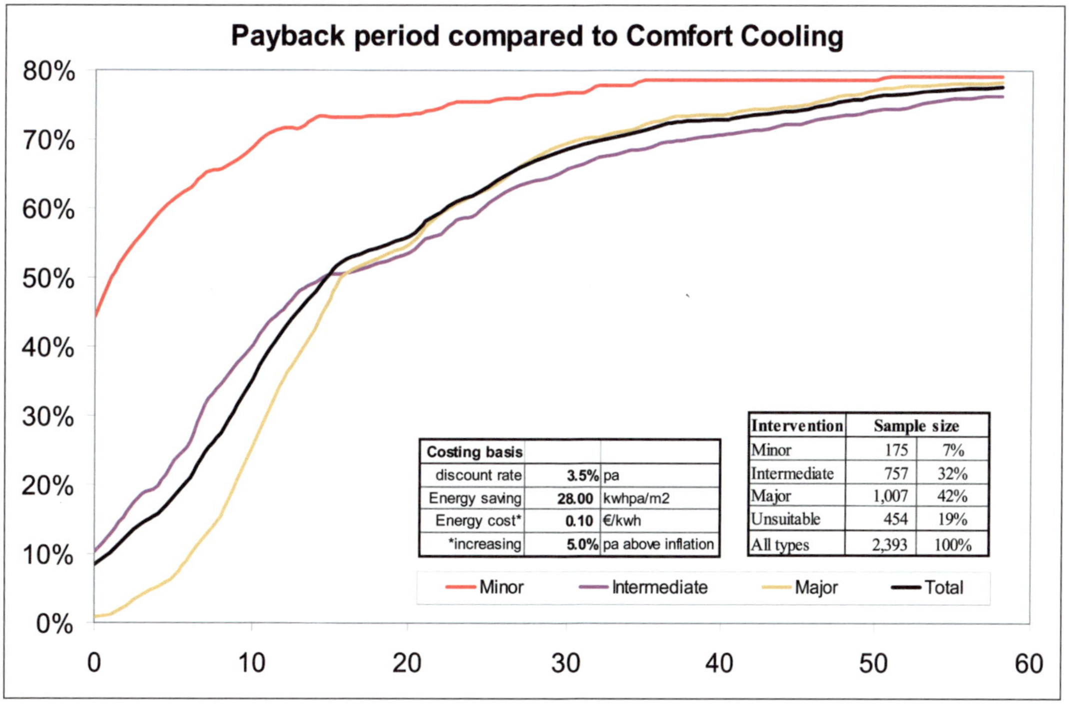
Payback period compared to Comfort Cooling
80%
70%
60%
50%
40%
30%
20%
10%
0%
0
10
20
30
40
50
60
Costing basis
discount rate 3.5% pa
Energy saving 28.00 kwhpa/m2
Energy cost* 0.10 €/kwh
*increasing 5.0% pa above inflation
Intervention Sample size
Minor 175 7%
Intermediate 757 32%
Major 1,007 42%
Unsuitable 454 19%
All types 2,393 100%
Minor
Intermediate
Major
Total

122

4.4 MARKET ASSESSMENT
Introduction

This section provides an assessment of the potential markets for PHDC in Europe, China and India. This assessment is based on a model initially developed to assess the market in southern Europe and then modified to show possible outcomes for all three regions. The original assessment was developed for the 'Altener' study[3], examining the potential for Passive Downdraught Evaporative Cooling in Europe, which included a review of potential market barriers to the uptake of PHDC.

Background to the model

It has been shown that PDEC (and PHDC) can be installed in suitable existing or new buildings. The Altener project studied the applicability and market potential of PDEC in southern Europe. This study looked at non-domestic buildings (mainly offices), although we will see that the potential application of PHDC in housing is also very significant.

In southern Europe there is a very large stock of existing buildings and a much smaller amount of new buildings adding to and replacing this stock. However, it is difficult to obtain a reliable estimate of the size of the stock. Therefore, estimates were made for each country based on population and economic statistics obtained from the CIA World Factbook[4], using space allowances typically applied in the UK and thought to be similar in Europe.

Morphological surveys were carried out in a number of European cities to establish the proportion of buildings that would be suited to installation of PDEC (as described in section 2.2 above). These factors were then applied to the stock estimates to give the total area of stock suitable for PDEC installation. Having determined the potentially suitable stock (the market), estimates of energy savings and CO2 reductions could be made based on earlier performance analysis.

PHDC Building Stock Methodology

It was equally difficult to obtain accurate stock data for India and China, so the methodology was refined to apply it to Asia as well as Europe, and to use it to estimate residential stock for which some types of PHDC would be suitable. Of course, the total theoretical market could be very different from what might be realistically achieved over time. The new model therefore provides best estimates. It is explained in detail, with the derivation of all the assumptions, on the attached CD together with the life cycle costing models described in Chapter 4. This model allows the user to enter a probable range rather than an exact figure against each parameter. Once these have all been aggregated the range for the potential market will be very wide reflecting the level of uncertainty regarding different parameters and the rate of 'take-up' of PHDC.

Key parameters		Source
1	Population millions	CIA World Factbook
2	% in suitable climatic region	User estimate
3	Labour force millions	CIA World Factbook
4	% in suitable climatic region	User estimate
5	% in services sector	CIA World Factbook
6	% of whom work in offices	User estimate
Area requirements		
8	Living space per person m2	User estimate
9	Average dwelling area m2	User estimate
10	Working space per person m2	User estimate
11	Average building area m2	User estimate
Existing Stock suitability		
13	Residential - substandard	User estimate
14	Residential - standard + quality	User estimate
15	Offices	User estimate
New stock suitability		
17	Residential	User estimate
18	Offices	User estimate
Electrical energy		
20	PHDC energy saving kWhpa/m2	Altener study
21	Total consumption million kWh	CIA World Factbook
22	Electricity price €/kWh	User estimate
23	CO2 emissions g/kWh	User estimate
24	CO2 target g/KWh	User estimate
25	over how many years?	User estimate
Annual changes		
27	Population growth	CIA World Factbook
28	Electricity consumption	User estimate
29	Electricity price	User estimate
30	Stock replacement residential	User estimate
31	- substandard	User estimate
32	Stock replacement commercial	User estimate
33	Service sector expansion	User estimate
34	PHDC final residential uptake	User estimate
35	over how many years?	User estimate
36	PHDC final commercial uptake	User estimate
37	over how many years?	User estimate

T 4.1

Table 4.1 - Key Parameters

Table 4.1 shows the key parameters required to be entered to set up the model. As many of these parameters as possible should be actual values, but most of them will be estimates. The actual figures used in this report for Europe, China and India are discussed in the following three sections. The CIA World Factbook provides the population (1), population growth (27), labour force (3), proportion working in the services sector (5) and electricity consumption (21) for most countries of the world. These are entered into the model as fixed quantities, but can also be entered as ranges. Numbers in brackets refer to parameter position in Table 8.1. The user must estimate all the other parameters with wider or narrower ranges depending on the degree of uncertainty.

The percentage in a suitable climatic region should be the percentage of the population; a poorer estimate would be the percentage by area of the country. Suitable climatic region might be described as hot and dry or hot and moderate humidity on existing maps, cold areas and high humidity areas should be excluded. We have allowed different percentages to be used for the whole population (for residential calculations) (2) and for the labour force (for commercial calculations) (4): for this study we have assumed that they are the same.

Although we have a good estimate for the percentage of the labour force working in the services sector, we need to estimate the proportion of these who work in offices (6).

We need to estimate the typical living space (8) occupied by each person (to calculate residential areas and the typical dwelling size (9) to estimate number of installations. Similarly, we need the average space requirements per office worker (10) and the average office building size (11).

When we consider the existing stock we need to estimate how much will actually be suitable for the installation of PHDC. We have some information for offices in Europe from the Altener study, but must assume those for India and China (15). For residential there are two considerations:

1. Is the stock sub-standard ?(13), a major issue in India and

we assume none of this is suitable for intervention, if not

2. What percentage of the remaining stock (14) would be capable of being adapted for PHDC?

For new residential (17) and new offices (18) we assume a higher proportion will be suitable for intervention, possibly all of it. From these inputs we can estimate a total stock area that potentially could incorporate PHDC and the number of installations this represents.

Potential Savings

The next stage is to assess the various savings we might make compared to the use of air conditioning. It is important to understand that this will be a hypothetical saving compared to the equivalent volume of air conditioning. The Altener study provided some estimates for achievable energy savings per m2 (20) in a number of European locations and average electricity costs (22) will allow us to estimate cash savings.

CO_2 emissions from electricity generation (23) vary around the world depending on the mix of fossil fuels (coal, oil, gas) and the proportion of clean sources (hydro, nuclear, wind, solar, etc.). As the model will be looking many years into the future we should also allow for reductions in emissions over time. The model allows us to enter a target range of emissions (24) and the number of years we are allowing for this to be achieved (25).

There are a number of other factors that are also likely to change over time. We have mentioned population growth above. Will energy consumption and therefore electricity generation increase and if so by how much per annum (28)? Will the price per unit increase? - It probably will by more than inflation to pay for alternative expensive generation and to deter excessive consumption. The model shows any costs in current (today's) values, only a percentage increase (29) over and above inflation should be used in the model.

We have given different percentage uptakes for PHDC in existing stock and new build and we need to know how much new build there is every year. There will be some to cater for a rising population but the majority will be to replace demolished existing buildings. If we can estimate the average life of buildings we can calculate the percentage needing replacement every year, this is usually lower for residential buildings (30) than for commercial buildings (32). We need to consider sub-standard stock separately; the model assumes that some stock will always be substandard but that it will be replaced at a different rate (31) than the general replacement rate, we presume that this would normally be at a higher rate, especially if national living standards are also rising. We have allowed for the possibility of a regular incremental expansion of the commercial sector (33). This may be higher in less developed countries and lower or non-existent in more developed countries.

The factors and constraints we have fed into the model so far allow the total potential market to be estimated for each year ahead, in other words complete market saturation. Realistically it will take time to build up to these levels and total saturation is unlikely to ever be achieved. The final inputs allow estimates for the levels of final saturation for residential (34) and commercial (36) and the number of years it might take to achieve this (35 and 37). The model assumes take up will follow an S curve with a slow start, speeding up and then slowing down again as saturation is approached.

Any forecast will be uncertain and there are many factors with large degrees of uncertainty in this model. We suggest that it should not be relied upon to look too far into the future!

Running the Stock Model

Once the Key Parameters described above have been entered onto the front sheet, the second produces the results for one country. A drop down box allows the country to be selected, a table similar to Table 3 above, displays the high, low ranges already entered, the mid range and a figure randomly chosen from within the range. A second drop down box allows a choice of which range will be used in the 4 standard charts provided with the model. An appraisal period needs to be set, this can be up to 100 years but note the caution above.

A final drop down box allows the output to be shown for residential buildings, commercial buildings or the total of both groups of buildings.

Below the charts the model sets out a table of outputs for every year in the appraisal period from which data can be copied and used for further analysis. The charts shown in the following country sections were built up from data extracted from the model.

Most of the parameters in the following country sections have been estimated by DL based on experience. We expect these outputs to show wide ranges between maximum and minimum estimates of potential and are not surprised by the following ranges averaged across all the forecasts:

Southern Europe	10:1
China	17:1
India	29:1

Review of the Current Market for PHDC

Application of the Building Stock model for Southern Europe (Portugal, Spain, Italy and Greece) indicates that the opportunity for residential electrical energy savings (and consequent reduction in carbon emissions) is approximately four times the potential savings achievable in the commercial building sector. Based on the assumptions built into the model, the value of residential electrical energy savings after five years of market adoption will be between 4-80million Euros/year. This represents up to 300,000 tonnes of CO_2/year, increasing to potentially 1million tonnes/year after 10 years. This is very significant – approximately equivalent to the total annual CO_2 emissions of 100,000 people living in Europe today.

This of course represents the potential following vigorous uptake of the technology. However, at the moment (May 2010), although there are a number of applications to non-domestic buildings in Europe, there is only one application to a house. It therefore appears that, although the total potential market is larger in the residential sector, the commercial or non-domestic sector is likely to provide the initial impetus for the take-up of PHDC as an alternative to conventional cooling technologies.

This pattern is likely to be similar in India and China. However, in India the application of PDEC in both domestic and non-domestic buildings is taking place already. Architects and engineers attending symposia on PHDC held in Delhi, Hyderabad and Ahmedabad in November 2009 showed a very high level of awareness and interest in the opportunity of adopting PHDC as part of their design practice. The building stock model for India shows that (as in Europe) the potential long term residential energy savings from PHDC are significantly greater than for the commercial sector. Based on assumptions of a 60% plus uptake over the long term in suitable locations in India, the residential savings in carbon emissions after five years of market adoption will be up to 300,000 tonnes of CO_2/year, and up to 1.3million tonnes/year after 10 years. This suggests a market which is similar in size to that in Europe in the first ten years, but in the long term the market is thought to be significantly larger (perhaps 2-300%) of the market in Europe.

In China, where today there is very little knowledge of PHDC, the building stock model suggests that the potential value of residential electrical energy savings attributable to PHDC will be similar to that in Europe (ie approx 70m Euros) after 5 years, but the potential saving in carbon emissions is much greater (rising to 3m tonnes after 10 years) due to differences in fuel mix. This level of take-up is only likely if there is vigorous promotion of PHDC in the regions of China where it is most applicable (ie the northwest).

Of course these projections must be taken with caution as there is substantial uncertainty associated with the take up of PHDC. Nevertheless, it is encouraging to note that there is already an established market in PHDC in India. This market is small, but judging from the number of new buildings in which it has been adopted in India, it also appears to be growing quite fast. The reasons for this may be related to the perception of PHDC as an affordable alternative to air conditioning. It may also relate to the willingness of architects and engineers in India to explore innovative solutions and a willingness among some clients to take more risks than their counterparts in Europe or the USA. It also almost certainly relates to the fact that the construction sector in both India and China are growing significantly faster than in Europe.

Market applicability in southern Europe

Table 4.2.1. Applying the new Stock Model to southern Europe.

* based on UK from carbonIndependant.org/

Key parameters		Source	S Europe from	S Europe to
1	Population millions	CIA World Factbook	120.10	
2	% in suitable climatic region	Altener study	100%	100%
3	Labour force millions	CIA World Factbook	58.51	
4	% in suitable climatic region	Altener study	100%	100%
5	% in services sector	CIA World Factbook	67.7%	
6	% of whom work in offices	DL estimate	50%	70%
Area requirements				
8	Living space per person m2	DL estimate	20	40
9	Average dwelling area m2	DL estimate	90	
10	Working space per person m2	DL estimate	15	20
11	Average building area m2	DL estimate	850	
Existing Stock suitability				
13	Residential - substandard	DL estimate	8%	3.0%
14	Residential - standard + quality	DL estimate	40%	50%
15	Offices	Altener study	75%	85%
New stock suitability				
17	Residential	DL estimate	50%	100%
18	Offices	DL estimate	90%	100%
Electrical energy				
20	PHDC energy saving kWhpa/m2	Altener study	25	30
21	Total consumption million kWh	CIA World Factbook	698,160	
22	Electricity price €/kWh	DL estimate	0.1200	
23	CO2 emissions g/kWh	*estimate	527	
24	CO2 target g/KWh	DL estimate	200	100
25	over how many years?	DL estimate	80	50
Annual changes				
27	Population growth	CIA World Factbook	0.037%	
28	Electricity consumption	DL estimate	0.50%	1.50%
29	Electricity price	DL estimate	1.50%	3.00%
30	Stock replacement residential	DL estimate	0.70%	2.00%
31	- substandard	DL estimate	1.00%	5.00%
32	Stock replacement commercial	DL estimate	1.5%	3.0%
33	Service sector expansion	DL estimate	0.00%	0.05%
34	PHDC final residential uptake	DL estimate	60%	80%
35	over how many years?	DL estimate	150	75
36	PHDC final commercial uptake	DL estimate	60%	80%
37	over how many years?	DL estimate	100	50

T 4.2.1

Chart 4.2.7. CO2 savings in Southern Europe.

Chart 7 data. CO2 savings tonnes per annum.

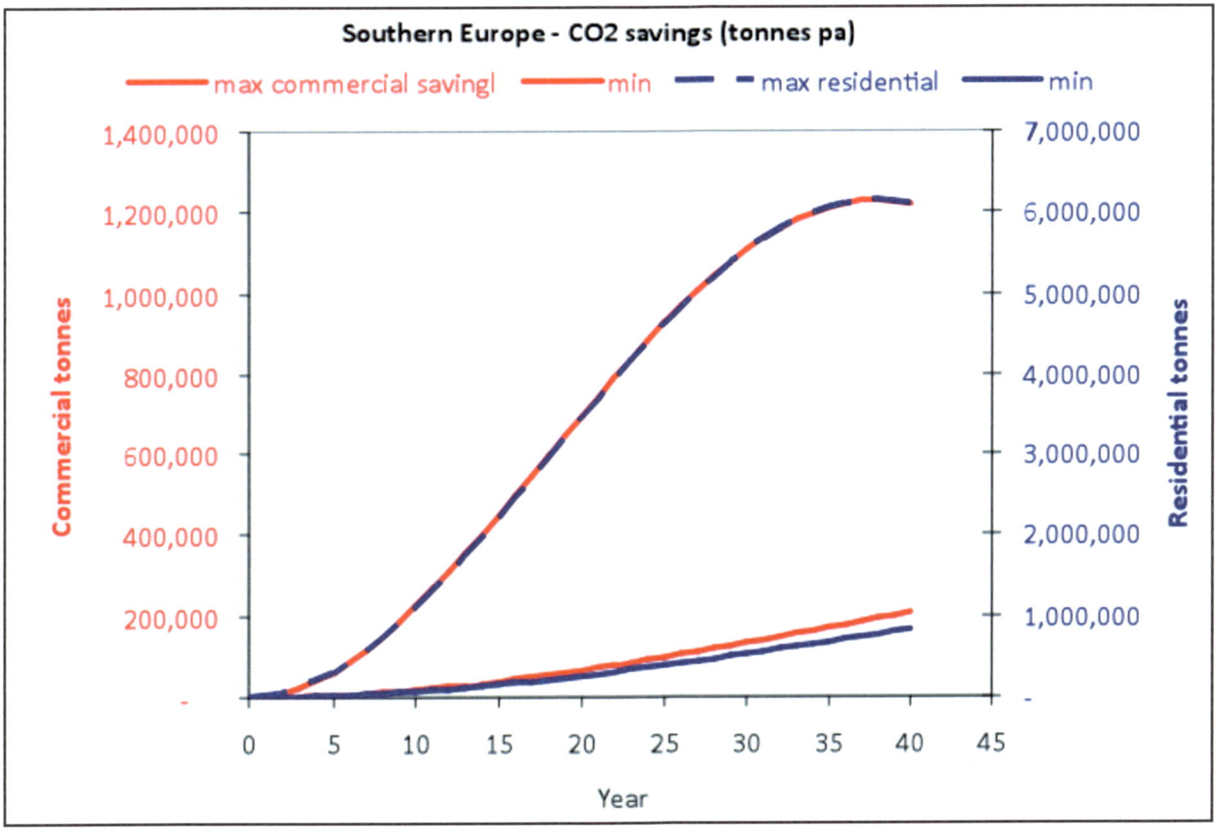

Chart 4.2.7

Sector	Estimate	Range	Year 1	5	10	20	40
Residential	CO2 saving tonnes	High	13,024	305,006	1,107,702	3,455,400	6,089,127
		Low	763	18,549	71,424	261,698	834,178
Commercial	CO2 saving tonnes	High	2,666	62,118	224,509	695,653	1,217,808
		Low	195	4,729	18,138	65,973	207,674

Chart 7 data

Key parameters		Source	China from	China to
1	Population millions	CIA World Factbook	1,338.61	
2	% in suitable climatic region	DL estimate	25%	35%
3	Labour force millions	CIA World Factbook	812.7	
4	% in suitable climatic region	DL estimate	25%	35%
5	% in services sector	CIA World Factbook	33.2%	
6	% of whom work in offices	DL estimate	40%	60%
Area requirements				
8	Living space per person m2	DL estimate	17	21
9	Average dwelling area m2	DL estimate	85	
10	Working space per person m2	DL estimate	12	20
11	Average building area m2	DL estimate	850	
Existing Stock suitability				
13	Residential - substandard	DL estimate	15%	5.0%
14	Residential - standard + quality	DL estimate	30%	45%
15	Offices	DL estimate	65%	95%
New stock suitability				
17	Residential	DL estimate	40%	90%
18	Offices	DL estimate	80%	100%
Electrical energy				
20	PHDC energy saving kWhpa/m2	Altener study	25	30
21	Total consumption million kWh	CIA World Factbook	3,438,000	
22	Electricity price €/kWh	*see note	0.0589	
23	CO2 emissions g/kWh	DL estimate	730	
24	CO2 target g/KWh	DL estimate	200	100
25	over how many years?	DL estimate	80	50
Annual changes				
27	Population growth	CIA World Factbook	0.655%	
28	Electricity consumption	**see note	3.00%	4.00%
29	Electricity price	DL estimate	1.50%	3.00%
30	Stock replacement residential	DL estimate	0.70%	2.00%
31	- substandard	DL estimate	1.00%	5.00%
32	Stock replacement commercial	DL estimate	1.5%	3.0%
33	Service sector expansion	DL estimate	1.00%	1.10%
34	PHDC final residential uptake	DL estimate	60%	80%
35	over how many years?	DL estimate	150	75
36	PHDC final commercial uptake	DL estimate	60%	80%
37	over how many years?	DL estimate	100	50

T 4.3.1

Market applicability in China

Table 4.3.1. Applying the new Stock Model to China.

* Xuan Huang dissertation 0.56Yen/kWh (9.5Yen=1€)

** http://www.iea.org/journalists/fastfacts.asp China consumption expected to double in 20 years

Chart 4.3.7. CO2 savings in China.

Chart 7 data. CO2 savings tonnes per annum.

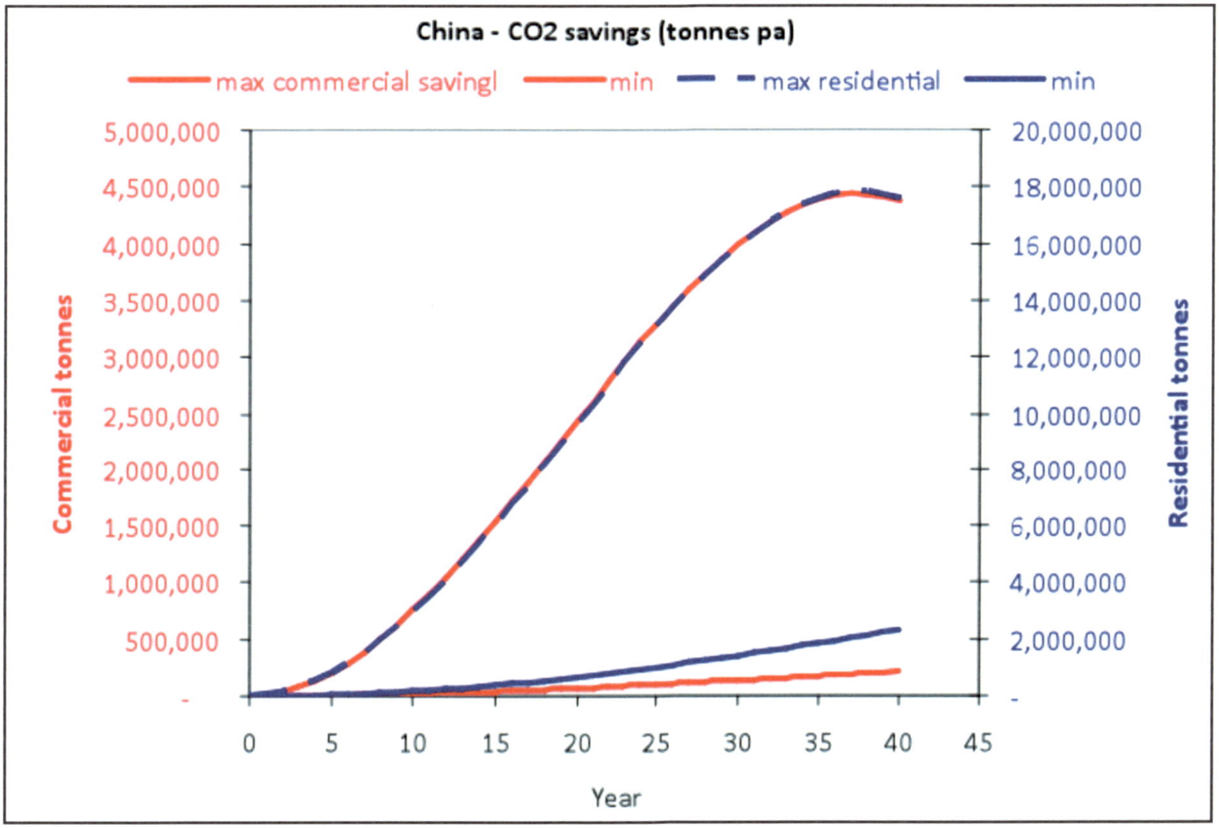

Chart 4.3.7

Sector	Estimate	Range	Year				
			1	5	10	20	40
Residential	CO2 saving tonnes	High	33,007	791,201	2,952,653	9,654,669	17,557,294
		Low	1,750	43,517	172,198	664,321	2,311,259
Commercial	CO2 saving tonnes	High	8,537	202,925	751,125	2,428,429	4,367,503
		Low	195	4,729	18,138	65,973	207,674

Chart 7 data

Key parameters		Source	India from	India to
1	Population millions	CIA World Factbook	1,156.90	
2	% in suitable climatic region	DL estimate	20%	30%
3	Labour force millions	CIA World Factbook	467	
4	% in suitable climatic region	DL estimate	20%	30%
5	% in services sector	CIA World Factbook	62.6%	
6	% of whom work in offices	DL estimate	40%	60%
Area requirements				
8	Living space per person m2	DL estimate	10	15
9	Average dwelling area m2	DL estimate	75	
10	Working space per person m2	DL estimate	8	20
11	Average building area m2	DL estimate	850	
Existing Stock suitability				
13	Residential - substandard	DL estimate	60%	40%
14	Residential - standard + quality	DL estimate	30%	45%
15	Offices	DL estimate	65%	95%
New stock suitability				
17	Residential	DL estimate	25%	75%
18	Offices	DL estimate	70%	100%
Electrical energy				
20	PHDC energy saving kWhpa/m2	Altener study	25	30
21	Total consumption million kWh	CIA World Factbook	568,000	
22	Electricity price €/kWh	DL estimate	0.0700	
23	CO2 emissions g/kWh	DL estimate	730	
24	CO2 target g/KWh	DL estimate	200	100
25	over how many years?	DL estimate	80	50
Annual changes				
27	Population growth	CIA World Factbook	0.072%	
28	Electricity consumption	*see note	3.00%	4.00%
29	Electricity price	DL estimate	1.50%	3%
30	Stock replacement residential	DL estimate	0.70%	2.00%
31	- substandard	DL estimate	1.00%	5.00%
32	Stock replacement commercial	DL estimate	1.5%	3.0%
33	Service sector expansion	DL estimate	0.00%	0.10%
34	PHDC final residential uptake	DL estimate	60%	80%
35	over how many years?	DL estimate	150	75
36	PHDC final commercial uptake	DL estimate	60%	80%
37	over how many years?	DL estimate	100	50

T 4.4.1

Market applicability in India

Table 4.4.1. Applying the new Stock Model to India.

* http://www.iea.org/journalists/fastfacts.asp India consumption expected to double in 20 years

Chart 4.4.7. CO2 savings in India.
Chart 7 data. CO2 savings tonnes per annum.

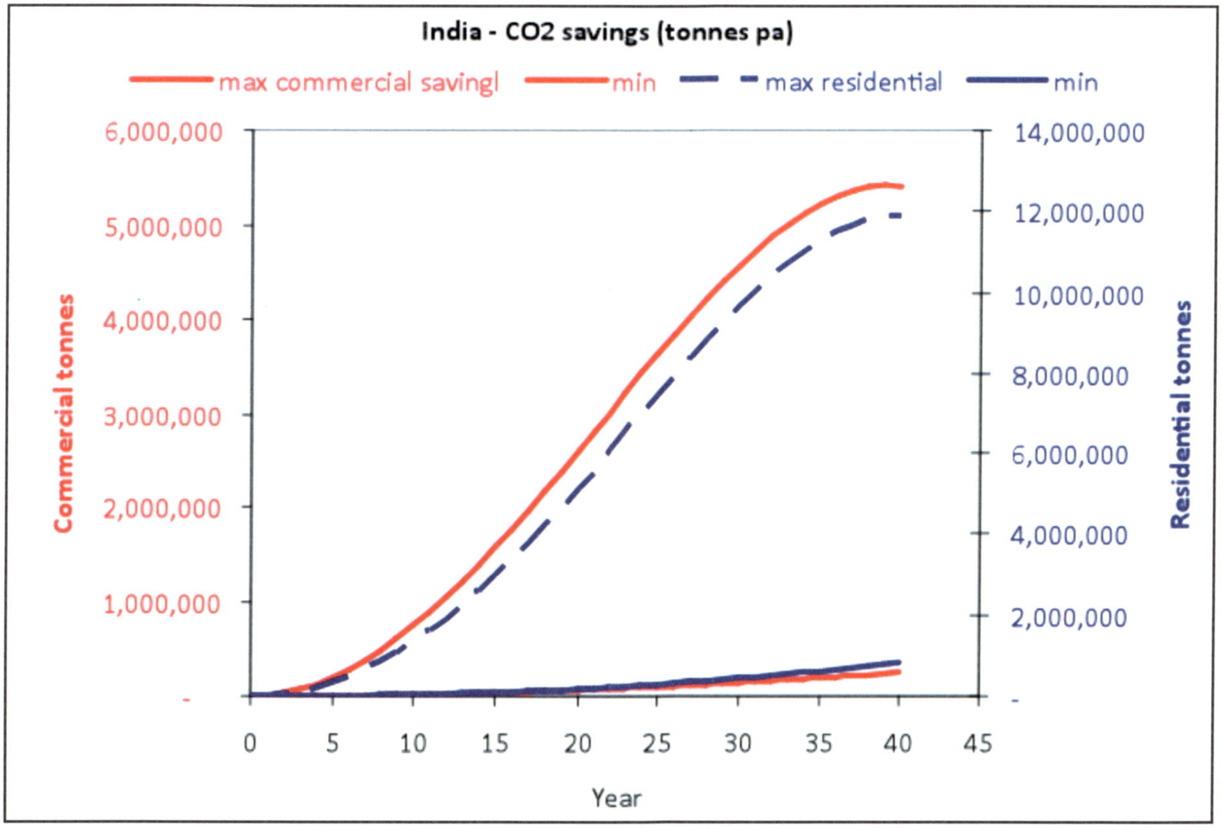

Chart 4.4.7

Sector	Estimate	Range	Year 1	Year 5	Year 10	Year 20	Year 40
Residential	CO2 saving tonnes	High	11,669	314,786	1,321,184	5,145,803	11,865,749
		Low	337	9,080	39,399	179,640	831,735
Commercial	CO2 saving tonnes	High	7,908	193,697	744,153	2,591,799	5,409,546
		Low	149	3,797	15,472	63,365	249,570

Chart 7 data

4.5 MARKET BARRIERS

As part of the Altener study a survey of the views of developers, designers and contractors suggested that market barriers in Europe could be categorised by six broad groups:

- Information and awareness barriers
- Technical barriers, including
  - comfort and ventilation standards
  - air pollution
  - noise levels[5]
- Tenure barriers
- Capacity barriers
- Economic and financial barriers
- Institutional and legal barriers

Of these, technical and regulatory barriers cannot be readily overcome. For some buildings for instance it would not be technically feasible to install PDEC, such as where the building is air conditioned or not in line for major refurbishments. And while regulation may not presently be a constraint it is possible that fire and air quality regulations for one reason or another disallow PDEC, and if so, this will become a major and serious barrier.

The principal economic barrier is willingness to pay for introduction or adaptation. The survey suggested that – relatively stringent - simple payback periods of 5-7 years were likely to be the maximum acceptable. Financial barriers could be addressed by institutional incentives such as grants, loans and tax incentives for the use of PDEC or disincentives for the use of alternatives.

Awareness is an important barrier for PDEC and its advocates to overcome. The evidence suggested that neither designers nor developers were well informed about PDEC, its attributes and its implications. The main route to improving awareness is information dissemination, and programmes would need to be designed and implemented for different audiences.

Regarding capacity constraints, it was recognized that

only after PDEC has been implemented on a significant basis will capacity be developed for design, installation and repair and maintenance (including spare parts).

But indeed the greatest barrier to natural cooling is likely to be the end consumer, who has become used to temperature controllable cooling deliverable to any point in a building at the touch of a switch. The performance of a natural ventilation and cooling system is more dependant on external weather; temperature, humidity and wind strength. Performance is likely to be an improvement in internal condition relative to external condition rather than the absolute condition achieved by air conditioning. It may take some time for consumers to accept this apparent reduction in performance although reactions from those working in current naturally ventilated buildings tends to be favourable.

In India and China it is possible that some of these categories may be less problematic. Potential consumers may be more open minded to new technologies and less conditioned to existing technologies.

CHAPTER 5
DESIGN SIZING – RULES OF THUMB

DESIGN SIZING RULES OF THUMB

Minimising the load: Internal and external heat gains

As with any cooling system, the 'size' or capacity of the system will be determined by the estimated cooling load. The first objective is to seek to minimise this load. The cooling load arises from internal and external heat gains which must be removed in order to prevent unacceptably high interior temperatures. (figs. 123 a - b - c)

THE FIRST OBJECTIVE IS TO SEEK TO MINIMIZE THE COOLING LOAD

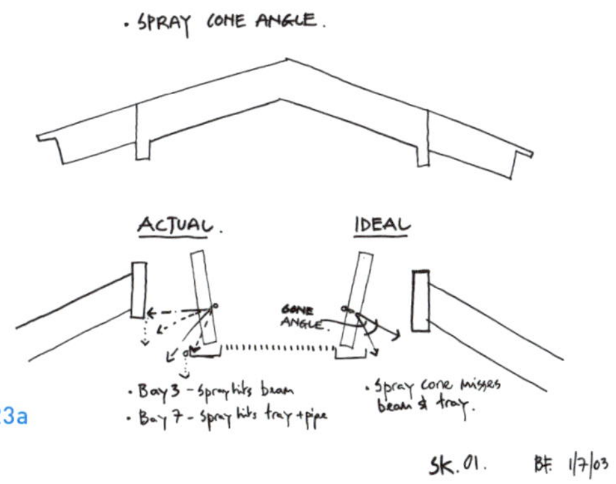

123a

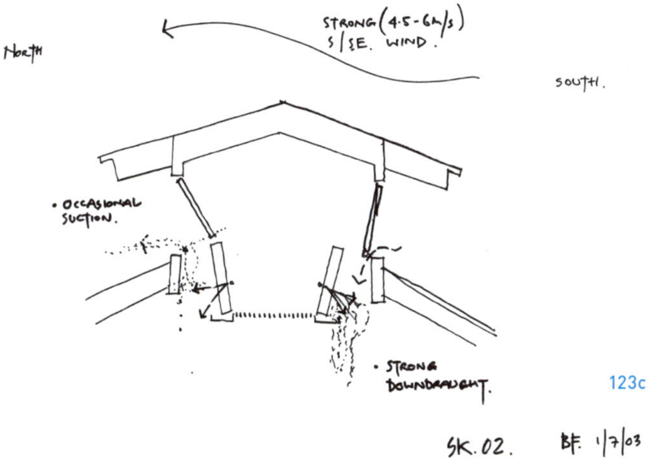

123c

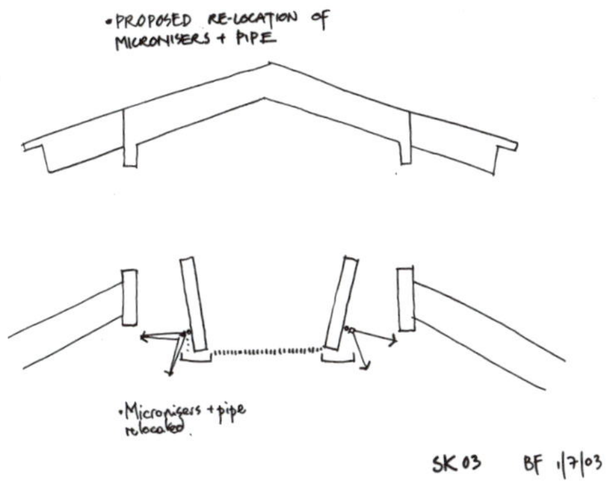

123b

123. (a-b-c) Diagrammatic Sketches of central rooflight/ventilator, cooling coils, misting nozzles and drip tray in central atrium of the Stock Exchange, Valletta, Malta.

5.1 INTERNAL HEAT GAINS

Internal heat gains are derived from people, lighting and equipment used within the building. Internal gains from people will depend on their activity and their metabolic rate, but typically are about 80W (sensible heat gain) for a sedentary adult. Heat gains from artificial lighting can hopefully be minimised by ensuring that most spaces are adequately daylit (i.e. do not require artificial lighting during daylight hours), while avoiding unwanted solar gains and glare. Heat gains from equipment will depend on activity and the efficiency of the appliances. Energy use related to domestic equipment has been rising in Europe and the USA in spite of improvements in efficiency. If we take the example of a small 75m2 four person apartment in Madrid, typical internal heat gains would be approximately 820 Watts. (figs. 124 a - b)

ENERGY USE RELATED TO DOMESTIC EQUIPMENT HAS BEEN RISING IN EUROPE AND THE USA IN SPITE OF IMPROVEMENTS IN EFFICIENCY

124. (a-b) Airflow strategy showing PDEC for summer daytime (top) and night ventilation (bottom) for Nottingham House for Solar Decathlon Europe 2010 Competition. (University of Nottingham).

124a

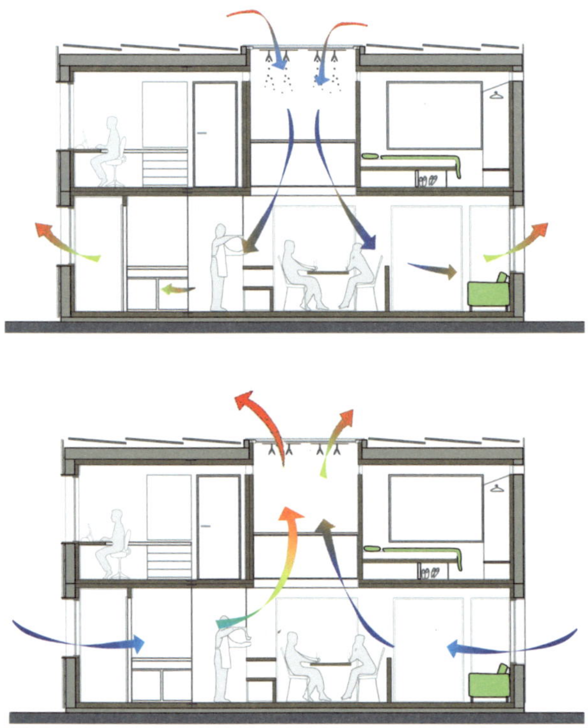

124b

Internal Heat Gains - 4 person flat

People: 4 x 80W = 320W
Lighting: allow 3W/m2 = 225W
Equipment: allow 3W/m2 = 225W
Total Internal Gains = 820W

External Heat Gains – 4 person flat Madrid

Heat gains through roof = 450W
Walls & glazing = 780W

Total fabric gains = 1230W

5.2 EXTERNAL HEAT GAINS

External heat gains arise from the prevailing ambient conditions (micro-climate) and are derived mainly from solar radiation (either directly or indirectly) and infiltration. Heat gains through the external opaque fabric will be directly related to the U-value of the construction, and also to the surface temperature which is related to the amount of solar radiation falling on the surface, or (in the absence of solar radiation) will relate to the temperature of surrounding surfaces and the ambient dry bulb temperature. External heat gains therefore vary significantly, and are related to the thermal capacitance and time-lag characteristics of the construction as well as the thermal transmittance (U-value). If we take the example of solar radiation falling on a flat roof in Madrid in the summer, this may reach as much as 1000W/m2. and, depending on the nature of the surface material, the temperature of the surface could reach 50-60°C. If the U-value of the roof was 0.2W/m$^{2°C}$,then (assuming an internal temperature of 25°C) the heat gain through the roof would be 6W/m2. If we take our topfloor apartment in Madrid, and assume the roof area is the same as the floor (75m2) then heat gains through the roof could be 450W. If we assume the external wall area of the apartment is 105 m2 (U-value 0.2W/m2°C for opaque areas) and 15m2 of the external wall area is fully shaded standard Low-E double glazing (U=1.4W/m2°C), then heat gains through the external wall could be approx 360 W (opaque walls) + 420 W (Glazing) (assuming an external wall surface temp of 45°C). Based on the foregoing assumptions, external heat gains would be 1230W and Total internal and external heat gains for the example apartment are 2050W.

EXTERNAL HEAT GAINS VARY SIGNIFICANTLY, AND ARE RELATED TO THE THERMAL CAPACITANCE AND TIME-LAG CHARACTERISTICS OF THE CONSTRUCTION AS WELL AS THE THERMAL TRANSMITTANCE (U-VALUE).

THE SUM OF THE INTERNAL AND EXTERNAL HEAT GAINS (CALCULATED FOR A HOT DAY IN SUMMER) IS THE COOLING LOAD WHICH HAS TO BE REMOVED BY DOWNDRAUGHT COOLING

The sum of the internal and external heat gains (calculated for a hot day in summer) is the cooling load which has to be removed by downdraught cooling. If the technique to be used is PDEC, then we can estimate the air volume flow rate required to remove this heat, based on an assumption about the temperature of the supply air arising from the evaporative cooling process. Of course this supply air temperature will vary with the external moisture content of the air and its dry bulb temperature. To be useful, PDEC must deliver air at a temperature which is below the desired internal air temperature. Givoni[1] suggests that the temperature achieved by direct evaporative cooling (the temp at the outlet of the tower) can typically be 80% of the dry bulb to wet bulb temperature depression. If, for example, the external dry bulb temperature is 40°C and the wet bulb temp is 20°C, then the tower outlet temperature may be assumed to be 24°C. This implies that room air temperature cannot be allowed to rise much above this. Typically, we can assume that room air dry bulb temperature will be 1 – 3°C above the supply air temp.

5.3 THERMAL COMFORT

THE HEAT BALANCE MODEL TREATS THE BUILDING OCCUPANT AS A PASSIVE RECIPIENT OF THERMAL STIMULI, WHEREAS THE ADAPTIVE COMFORT MODEL TREATS THE OCCUPANT AS AN ACTIVE PARTICIPANT, INTERACTING AND ADJUSTING TO THEIR ENVIRONMENT

The objective of course is to achieve thermally comfortable conditions for the occupants of the building. Two different (and competing) models of thermal comfort have developed over the last 30 years: a heat balance model, based on laboratory studies (largely undertaken by Ole Fanger[2]), and an adaptive comfort model, based on field studies, initially put forward by Humphreys and Nicol[3]. The heat balance model treats the building occupant as a passive recipient of thermal stimuli, whereas the adaptive comfort model treats the occupant as an active participant, interacting and adjusting to their environment. A large number of field studies have shown that contextual factors have significant influence on perceptions of the thermal environment, and that the adaptive model provides a more accurate prediction of building occupant perceptions and responses than the heat balance model[4].

ADAPTIVE COMFORT THEORY HAS GAINED WIDESPREAD ACCEPTANCE AS A REASONABLE BASIS FOR DEFINING THE ACCEPTABLE UPPER LIMIT OF TEMPERATURE IN BUILDINGS IN DIFFERENT PARTS OF THE WORLD.

Adaptive comfort theory[5] has gained widespread acceptance as a reasonable basis for defining the acceptable upper limit of temperature in buildings in different parts of the world. Indeed the adaptive model has now been incorporated into major comfort standards around the world, including the European EN 15251 standard, the US ASHRAE 55 – 2004, and the UK CIBSE Comfort Standard, 2005[6].

Determining the air volume flow rate required to remove the heat gains

Returning to our example of the apartment in Madrid, we now wish to determine the air volume flow rate required to remove the estimated total heat gains. The required cooling (Qc) is a function of the volumetric specific heat of air (assumed to be 1200J/m3), the air change rate/hour (N), the volume of the space (V), and the temperature difference between the PDEC supply air temp(Tt) and the max desired room temp (Tr). Since we know the cooling required the total heat gains to the space), the air volume flow rate required to remove the heat gains, can be derived from the expressions below:

Cooling & Air Volume Flow Rate

$$\mathbf{Qc}\ (\text{Watts}) = \frac{1200 \times \mathbf{N} \times \mathbf{V} \times (\mathbf{Tr}\text{-}\mathbf{Tt})}{3600}$$

and,

$$\mathbf{N}\ (\text{ac/hr}) = \frac{\mathbf{Qc}}{0.33 \times \mathbf{V} \times (\mathbf{Tr}\text{-}\mathbf{Tt})}$$

Area of Ventilation Opening

$$\text{Area } \mathbf{A}\ (m^2) = \frac{\mathbf{Qv}}{\mathbf{Cd} \times \mathbf{v}}$$

Where: **Cd** = discharge coefficient = 0.6

Qv = Volumetric air flow rate (m3/s)

v = air velocity (m/s)

The total heat gains (cooling load) is 2050 W (determined above), and N (the air change rate required) can be determined from the equation. For this example we find N to be 9.1 air changes or 0.57m3/s.

> IF WE KNOW THE VOLUME FLOW RATE REQUIRED, THEN WE CAN SIZE THE VENTILATION OPENINGS REQUIRED BY ASSUMING AN AVERAGE VELOCITY THROUGH THE OPENINGS

If we know the volume flow rate required, then we can size the ventilation openings required by assuming an average velocity through the openings. Typically with natural ventilation, air velocities through openings are rarely much more than 0.3m/s. For the apartment in Madrid this suggests we need 1.14m2 free area for both inlet and outlet openings.

The method outlined above can give you a very rough 'first stab' at opening sizes required, and can be used to 'pre-size' the system and its components in the 'PHDC Airflow' Software (Appendix C). The dimensions and characteristics of the systems can then subsequently be modified after obtaining the results from the program. (see also section 5.4 below)

Sizing the Downdraught Cooling Tower and Vent Openings

We have said before that passive downdraught cooling is a form of natural ventilation, driven by either buoyancy (stack effect) or wind or a combination of the two. If we take the case of buoyancy or stack effect first, the volume flow rate is a function of the area of inlet and outlet openings (A1, A2...), the height between them (H1, H2...), and the temperature difference between supply air temperature and room air temperature (Tr – Tt).

With stack driven airflow, as the supply air and the room air temperatures equalize, the driving pressures will approach zero, and no matter how large the opening, the flows will be very small. (fig. 125) shows that the cooling effect is very small at low temperature differences.

> THE HEIGHT DIFFERENCE BETWEEN INLET AND OUTLET OPENINGS HAS A FUNDAMENTAL INFLUENCE ON THE AIR FLOW RATE ACHIEVED, AND ON THE AREA OF OPENINGS REQUIRED TO ACHIEVE THE REQUIRED FLOW RATE.

We have already said that typically, we can assume that room air dry bulb temperature will be 1 – 3°C above the supply air temperature, and therefore large air change rates are generally required to achieve significant cooling of the building. Cunningham & Thompson reported air change rates of up to 30/hr for their experimental building in Arizona[8] under buoyancy forces alone, and at the TRC in Ahmedabad (see Chapter 3), air change rates of 6 -12/hr were recorded[9]. Experience suggests therefore that buoyancy driven cooling can achieve the large air volume flow rates required.

The height difference between inlet and outlet openings has a fundamental influence on the air flow rate achieved, and on the area of openings required to achieve the required flow rate. In a PDEC building, openings higher up the PDEC tower will need to be larger than openings lower down the tower if an even flow rate is to be achieved on each floor of the building. (fig. 126)

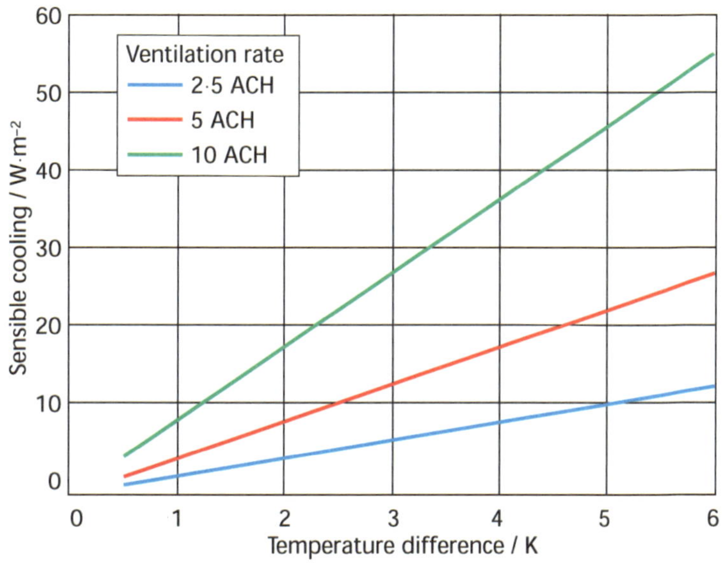

125

125. Effect of temperature difference and air change rate on sensible cooling capacity of ventilation air. (height of space taken as 2.75m)

126. Summary rules of thumb for sizing openings for PDEC strategy.

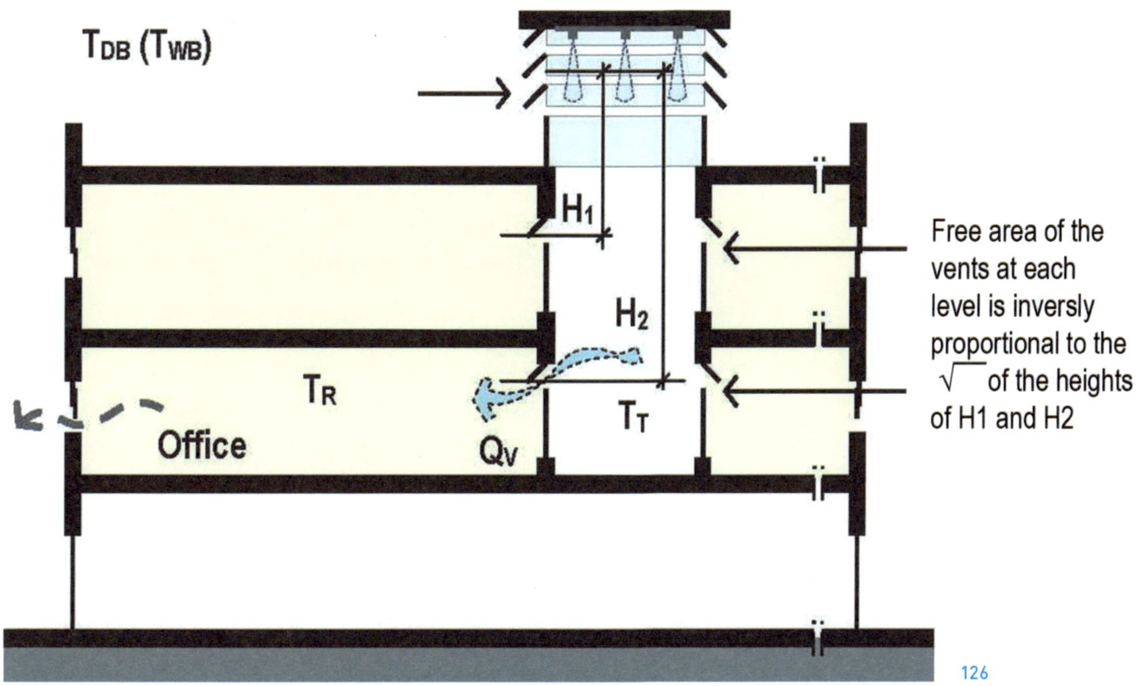

126

5.4 ESTIMATING VENT OPENING SIZES

The area A of each opening required to give a flow rate q for a specified value of h is: (fig. 127)

127 $$A = \frac{q}{C_d}\sqrt{\frac{(T_1+273)}{\Delta T\, g\, h}}$$

Where A is the area of each opening (m2), q is the air flow rate (m3/s), Cd is the discharge coefficient, Ti is the internal (room) temperature (°C), ΔT is the difference between the internal and supply air temperatures (K), g is the gravitational force per unit mass m/s2 and h is the height between openings (m). The discharge coefficient for a rectangular opening is usually taken as 0.6, and g the gravitational constant = (9.8m/s2)

So, continuing with the example of the Madrid apartment

q = 0.57m3/s

Ti = 27 C

ΔT = 3 K

H = 10m

A = 0.95 m3/s * 1.009 s2/m2 = 0.958m2

For preliminary calculations, it is assumed that the total inlet area will equal the total outlet area, and that the cross-sectional area of the supply tower will also be equal to the effective free area of the inlet area (Of course it may be larger if it is serving another function).

For each opening the pressure differences will of course be different, and so the building geometry has a profound impact on both the airflow pattern and the volume flow rate through separate floors. This further underlines the importance of defining a viable airflow strategy as described above (see section 1.4).

127. Equation from CIBSE AM10 p 45.

128. (a) Cool Tower Exit Air Temperature as a function of Wet-Bulb Depression and Outdoor Dry-Bulb Air Temperature.

128. (b) Recommended cool tower height and wetted pad area as a function of required airflow rate and wet bulb depression.

Summary "Rules of Thumb" for PDEC Performance

Temperature depression

Tt = Tdb - 0.8 (Tdb - Twb)

Where: **Tt** = Tower delivery temperature

Tdb = ambient dry bulb temperature

Twb = ambient wet bulb temperature

Assume *Tr = Tt + 3°C*

Cooling Achieved *(Qc) = 0.33 Qv (Tr - Tt)*

Design Sizing for Cool Towers

Whatever type of downdraught cooling delivery system is chosen, the viability of the envisaged airflow path, and hence the likely cooling performance which can be achieved, must be assessed as part of the design process. While a dynamic thermal and airflow model of the whole building is therefore likely to be required (certainly for large or complex buildings), for preliminary design sizing of the cool tower elements a simple steady state model can be applied. Thompson et al[10] suggest a method of relating the height of the cool tower and the area and 'efficiency' of the evaporative cooling matrix or pad, to the velocity (and thus volume flow rate) of the air within the tower. Givoni[11] has developed similar expressions which he has used to develop a series of monograms[12] from which exit air temperatures and air flow rates can be derived depending on tower height and pad area. Kwok & Grondzik[13] also present a cool tower design procedure, as follows:

WHATEVER TYPE OF DOWNDRAUGHT COOLING DELIVERY SYSTEM IS CHOSEN, THE VIABILITY OF THE ENVISAGED AIRFLOW PATH, AND HENCE THE LIKELY COOLING PERFORMANCE WHICH CAN BE ACHIEVED, MUST BE ASSESSED AS PART OF THE DESIGN PROCESS.

Design Procedure for 'Cool-Tower' sizing

1. Establish design conditions. Find design ambient dry bulb(DB) and mean coincident wet bulb (WB) air temperatures for the hottest time of the year for the building site, and derive the wet bulb depression (DB – WB).

2. Find approx exit air temperature by reference to. (fig. 128 a) The exit air temp needs to be at least 1-2°C below the acceptable max internal air temperature.

3. Determine required airflow rate need to remove the space cooling load from equation. (fig. 127)

4. Determine tower height and area of wetted pads required to provide this air volume flow rate from. (fig. 128 b)

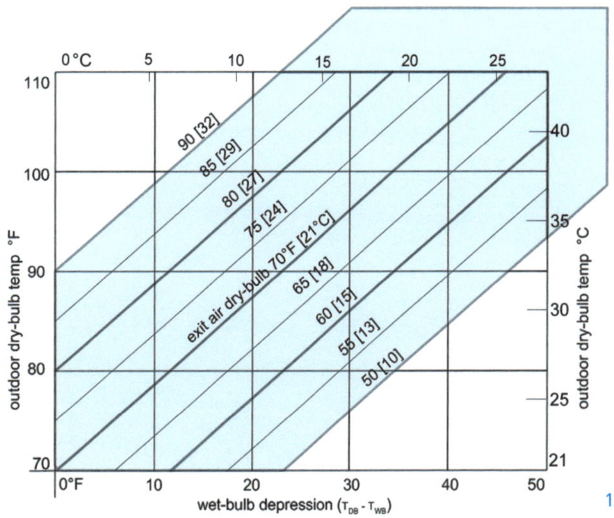

128a

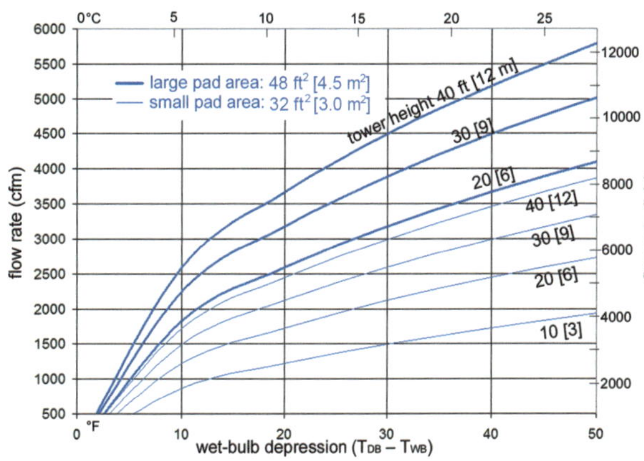

128b

Sizing the Porous Ceramic System

Given the appropriate climatic conditions, the technical applicability of a porous ceramic cooling system is dependent on the ratio between useful perimeter walls and required width of the porous ceramic panels. The area of panels required to meet the space cooling demand can be obtained from the performance chart (fig. 129) developed as part of the Evapcool project[14]. The chart gives the specific cooling power of different heights of the porous ceramic panels expressed as a function of the ambient wet bulb temperature depression. By entering the temperature depression at the peak hour of a typical summer day, it is possible to identify the system's specific cooling power for each system's height. From the specific cooling power it is possible to derive, the width of the panels to be integrated into the building envelope.

For a dry-wet bulb temperature depression of 10K, for example, the specific cooling power of the porous ceramic panels ranges between 40 and 68W per square meter of surface area and temperature differential K. The variation in specific cooling output depends on the height of the system. Shorter systems (h=0.2m) offer greater specific cooling power due to increased efficiency; however, a greater width may be necessary to obtain the total cooling capacity required. On the other hand, taller systems (h=1m), which offer smaller efficiency since the evaporative cooling effect is diminished by the saturation of the air along the system's height, can provide the required total cooling capacity with less width, which is easier to integrate into the envelope.

Dynamic thermal modelling and CFD analysis for a case study apartment in Seville[15-16] suggested that a porous ceramic system could provide temperatures below 28°C for 75% of the occupied time (assuming good solar control, exposed thermal mass and night time natural ventilation).

Summary of performance opportunities and risks

One of the inherent risks of most Cool Tower designs is that, a combination of solar gains and wind pressure can result in significant losses to ambient, resulting in higher levels of water use than with misting systems. Additionally, depending on the thickness of the cellulose pads, droplets of water can be carried into the building. This can be overcome by specifying thicker pads, but this also results in further increasing the resistance to airflow (already a significant factor). The general conclusion must be that, unless water quality is very poor, misting nozzles are probably the preferable option, providing the most efficient method of inducing downdraught evaporative cooling.

THE GENERAL CONCLUSION MUST BE THAT, UNLESS WATER QUALITY IS VERY POOR, MISTING NOZZLES ARE PROBABLY THE PREFERABLE OPTION, PROVIDING THE MOST EFFICIENT METHOD OF INDUCING DOWNDRAUGHT EVAPORATIVE COOLING.

129. Nomogram to show cooling potential of porous ceramic evaporators as a function of the temperature difference and height of the evaporators.

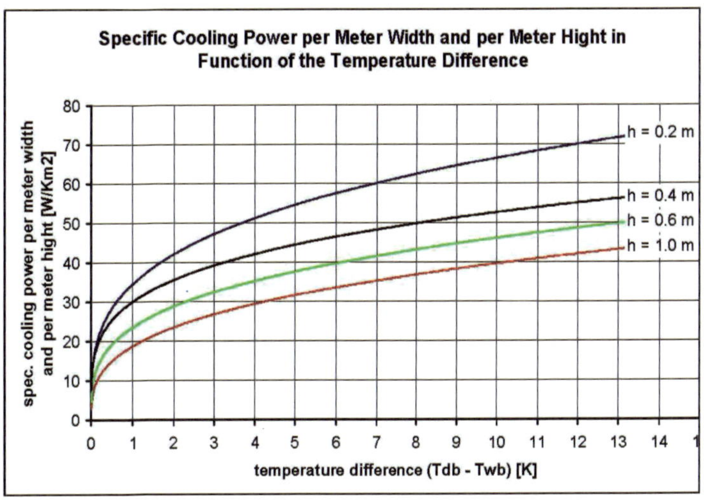

129

Sp

CHAPTER 6

PERFORMANCE ASSESSMENT

6.1 INTRODUCTION

As we have seen in section 1.2, natural cooling is the dissipation of heat from buildings to an environmental heat sink, like air, water, sky or ground. These are the traditional cooling systems that dominated architectural design in hot climates prior the availability of low-cost electricity.

This chapter describes the background to the development of two simplified software tools (created by AICIA), which are designed to assist architects and engineers in taking forward preliminary designs for the application of PHDC to a building.

In Passive Downdraught Evaporative Cooling, during the evaporative process, the transfer of heat energy from the external supply air to water supports the phase change process and so reduces the temperature of the ambient air that finally is driven into the building.

The performance of a PHDC application depends on the following:

- The climate, which provides the availability of the heat sink in terms of its thermal level -wet bulb temperature- and its variability throughout the year and on a daily basis.
- The cooling requirements of the building (absolute values and load profiles). The combination of 1 and 2, gives the extent in which the requirements can be covered by the cooling technique.
- The efficiency of the technology used to reduce the temperature from the ambient air to the wet bulb temperature.

The electricity consumption (in compressors, fans and/or pumps) linked to the use of PHDC techniques is very low compared to that required for conventional cooling systems.

THIS CHAPTER DESCRIBES THE BACKGROUND TO THE DEVELOPMENT OF TWO SIMPLIFIED SOFTWARE TOOLS (CREATED BY AICIA), WHICH ARE DESIGNED TO ASSIST ARCHITECTS AND ENGINEERS IN TAKING FORWARD PRELIMINARY DESIGNS FOR THE APPLICATION OF PHDC TO A BUILDING.

Limitations on the applicability of PDEC

It is evident from experience in case study buildings (see chapter 3 and Appendix A) that in certain locations and/or times of the year, evaporative cooling can meet the full cooling requirement. However, at certain times, the cooling produced may not be sufficient to cover 100% of the cooling requirements (depending on building type, occupancy pattern and location). In these circumstances, an auxiliary system has to be operated in conjunction with PDEC to meet the cooling load. Also, it can happen that at or near the peak cooling period, the PDEC techniques can not be operated at all, because the outdoor relative humidity is too high or because the temperature of the supply air is above the desired indoor temperature. During these periods, the cooling requirements have to be totally covered by auxiliary equipment. This has resulted in the development of hybrid downdraught systems.

The second case can limit the application of PHDC. In some cases this auxiliary equipment may be conventional fan-coil units or centralised air-conditioning. However, where it is possible to use chilled water cooling coils to provide supplementary buoyancy driven cooling when conditions are not appropriate for evaporative cooling, this will avoid the need for fans, air handling units, ductwork etc, reducing both capital cost and the energy cost of the fans. It is therefore desirable that the auxiliary cooling system mirnics PDEC in promoting naturally driven circulation of air within the building or space.

Where it is not possible to use buoyancy driven auxiliary cooling, only a significant reduction of the running cost will make the use of PHDC techniques viable from a cost/benefit point of view, since PHDC has a similar initial cost compared with the conventional system. Higher running costs are linked to non-residential buildings, so that for residential buildings the PHDC techniques are likely to be feasible if no auxiliary cooling is required. (Cost benefit analysis is dealt with in Chapter 4)

6.1.1 SYSTEMS DESCRIPTION AND PERFORMANCE

Having established the applicability of PHDC in a particular location, the type of evaporative cooling system can be chosen. As previously described, the three main alternatives consist of an air supply tower incorporating either:

1. a water spray system, or
2. a water sprinkler system, or
3. a wetted porous pad

The water spray system is obtained using rows of micronisers (nozzles which produce an artificial fog by injecting water at high pressure through minute orifices). In this case the water droplets are evaporated totally before the end of the tower.

THE USE OF MICRONISERS LEADS TO A REDUCTION IN THE WATER CONSUMPTION AND A HIGHER DOWNDRAUGHT EFFECT IN COMPARISON WITH THE WATER SPRINKLER SYSTEM.

The water sprinkler system produces water droplets that are not evaporated totally, so it requires recovering and recirculating the excess of water.

The wet pad system consists of a porous media that can be made from cellulose or clay. Its thickness varies typically between 5 cm to 10 cm. The cell size varies between 1 mm to 2 cm.

The use of micronisers produces a much better control of the system and a significant reduction of the pressure losses in comparison with the porous pad system. Also the use of micronisers leads to a reduction in the water consumption and a higher downdraught effect in comparison with the water sprinkler system.

The psychrometric chart (fig. 130) shows the conditions of the ambient air (E0), the outlet conditions of air which is cooled by the direct evaporative cooling system (E1) and the outlet conditions of air which cooled by the indirect evaporative cooling system (E2).

130. The process of direct and indirect evaporative cooling as shown on the psychrometric chart.

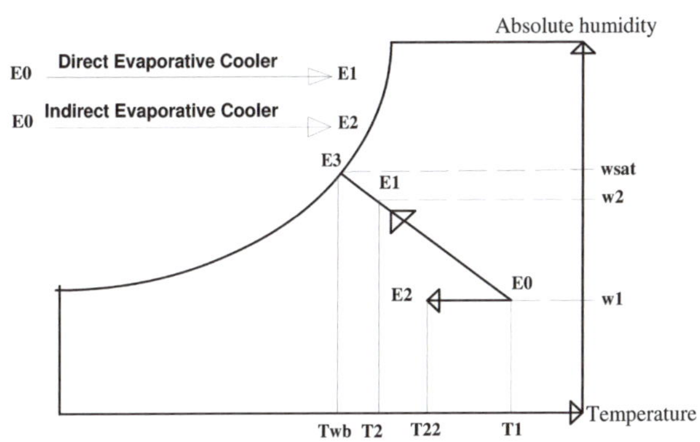

130

6.1.2 DESIGN AND SIZING OPTIONS

The previous chapter has given general guidance and 'rules of thumb' for preliminary sizing of the main elements of PHDC systems. However, once basic feasibility has been established, and an outline design has been defined, a more detailed analysis is required in order to progress the design and give more confidence in predicting performance. The two software tools which have been developed as part of the PHDC project should be considered for use before CFD or more detailed dynamic thermal modelling is applied. Depending on the nature of the project and its location, the design team may feel confident to proceed on the basis of using these tools, but for large,

> ONCE AN OUTLINE DESIGN HAS BEEN DEFINED, A MORE DETAILED ANALYSIS IS REQUIRED IN ORDER TO PROGRESS THE DESIGN

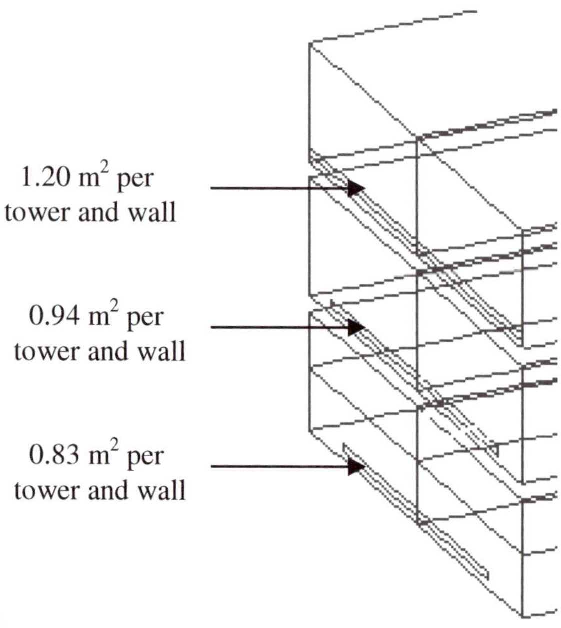

131

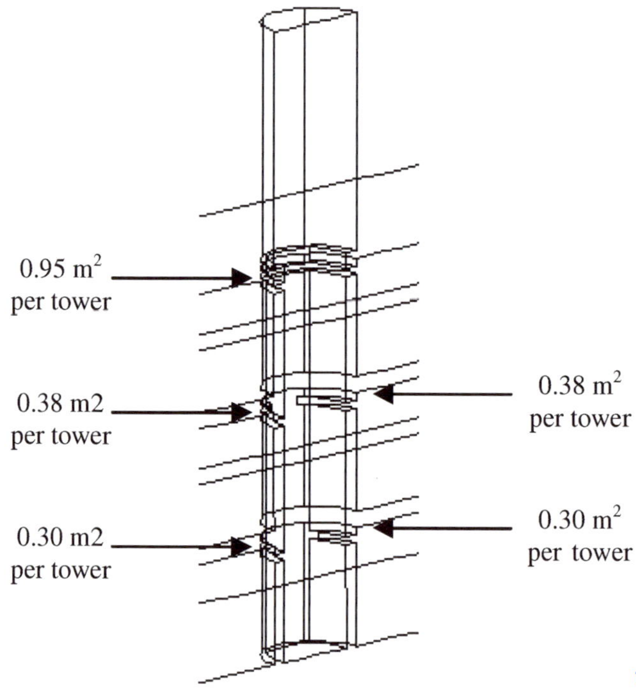

132

complex or very prestigious projects Computational Fluid Dynamic software may be required.

The applicability of evaporative systems in regions other than arid is closely linked to the use of these systems in combination with other techniques or strategies.

The combination of the PHDC Air Flow and Energy & Comfort tool is very useful to do some initial simulations changing any of the parameters concerning the building, climate and PHDC system. It is also very useful to analyze the results obtained before carrying out a detailed simulation using CFD.

The previous section illustrated that the optimization of the tower design involves many connected issues:

1. Height of the head.
2. Shape, size and position of the connections between the tower and the building (inlets).
3. Shape, size and position of the connections between the building and the outside (outlets).
4. Relative position of inlets and outlets.
5. Layout of the towers in the building plan.

This section illustrates some conclusions regarding these issues that have been obtained during the development of the investigations carried out in previous EC projects.

131. Outlet areas required at the different floor levels within the proposed office building for Catania, Italy.

132. Inlet areas required at the different floor levels within the proposed office building for Catania, Italy.

Shape, size and position of the connections building/outdoor (outlets).

It is very important to tune the outlet areas in order to avoid the air getting blocked inside the building as this situation will lead to a block of the whole system and no air movement will be obtained.

There are four possibilities:

1. Absence of exhaust tower.
2. Presence of exhaust tower, and in this case:
3. Absence of auxiliary exhaust fan, or
4. Presence of auxiliary exhaust fan.

If an exhaust tower is not included, the air has to leave the building through openings in the facade, which must be protected from wind effects by external baffles, for example, by means of a 'double', or ventilated façade. The internal layer may be a glazed facade with openings for the PDEC outlet. The external layer may be glazed or opaque, depending on the need for shading, daylighting etc. It is clear that the PDEC system needs to be carefully integrated within the building as a whole.

> IF AN EXHAUST TOWER IS FORESEEN BUT NO FANS ARE GOING TO BE INSTALLED IT IS VERY IMPORTANT TO STUDY THE THERMAL BUOYANCY FORCES INSIDE THE TOWER IN ORDER TO GUARANTEE THE AIR MOVEMENT IN EVERY REQUIRED MOMENT

It is recommended that the area of the exhaust openings is studied in detail both to remove internal heat gains and in order to attempt to equalize the air flow rates on each floor; this can be done using the PHDC Air Flow software that has been developed in the framework of this project. The work of detailed performance prediction and precise design was undertaken in relation to the Catania building developed by MCA (Mario Cucinella Architects – see Chapter 2) in the framework of PDEC project.

If an exhaust tower is foreseen but no fans are going to be installed it is very important to study the thermal buoyancy forces inside the tower in order to guarantee the air movement in every required moment. This has been done previously for example in the TRC building in Ahmedabad (described in chapter 3) which has been working very satisfactorily every summer for the last ten years. For this purpose we have developed the software tool PHDC Air Flow.

Installing an exhaust fan allows the air flow to be controlled depending on the cooling energy demanded and the efficiency of the evaporation in each moment. It also could guarantee the air movement unless there is a power cut, maintenance or control problems. The presence or absence of exhaust fan is a possibility that can be simulated using the PHDC Air Flow software.

Shape, size and position of the connections between tower and building (inlets).

Different inlet areas are required in order to equalize the air flow rate in each floor. Inlet areas generally need to be larger on upper floors and smaller on lower floors because the pressure differences within the inlet stack increase with stack height. If the size or the characteristics of pressure losses of the grilles are not changed between different floors the air flow could be very different from one floor to another, resulting in differences of temperature and comfort perception.

Figure 132 shows the change of opening sizes at each floor level for the example building.

The orientation of the inlet area is a key issue also. Inlet area must be distributed around the tower pointing to the outlet, but best results are achieved when the entire area is widespread throughout the perimeter of the tower.

Relative position of inlet and outlet

For short distances between inlet and outlet, short-circuit risk makes the top-bottom layout better than the top-top one. This is caused by the tendency of the cooled air to leave the office without heat exchange when short distances and inlet and outlet are placed adjacent to each other (i.e. small offices and top-top).

For example, figure 133 shows the streamlines in a section of a room where the air inlet is top on the right and air outlet is bottom on the left. There we can see that the air gets in contact with walls 1 and 2 but not so much with 3 and 4. If a short circuit exists, the air would get in contact only with one wall –the 3 if a top-top configuration is used. More information regarding this topic can be found in Salmeron et al[1], where the flow pattern effects on night cooling ventilation are analysed.

For deeper spaces, it is preferable to adopt top inlet-bottom outlet since short-circuit risk has disappeared and by using this configuration incoming fluid must follow a longer path to leave the enclosure. This obviously improves heat exchange.

133

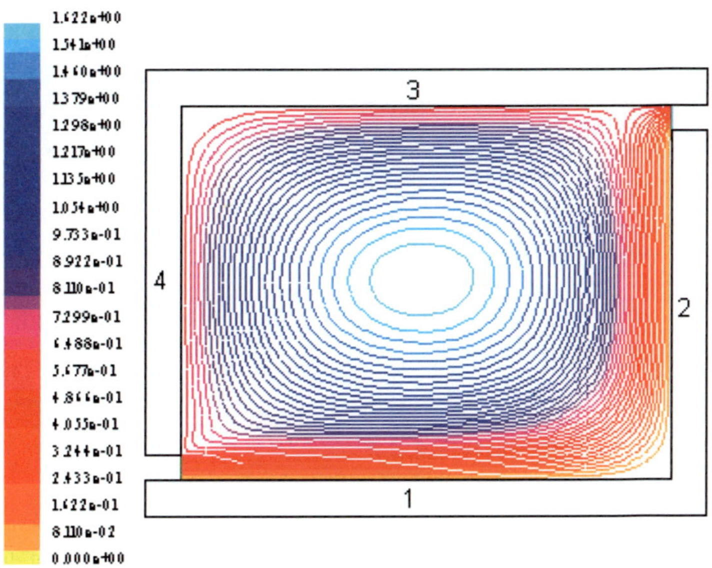

133. Streamline for a top-bottom configuration obtained using Fluent © software. (AICIA).

Layout of the towers in the building plan.

Figure 134 shows four different configuration of inlet tower in a 150m2 area of a notional office building. The following tables show the results obtained for all the simulations solved with the next assumptions:

1. Internal heat generation of 30 W/m2, this is the heat generation in the office due to people and equipment.
2. Inlet velocity values of 1m/s, 2m/s and 3m/s
3. Adiabatic walls
4. Constant properties of the fluid (air)

134. PDEC shaft layout options considered for floors within the proposed office building for Catania, Italy.

134

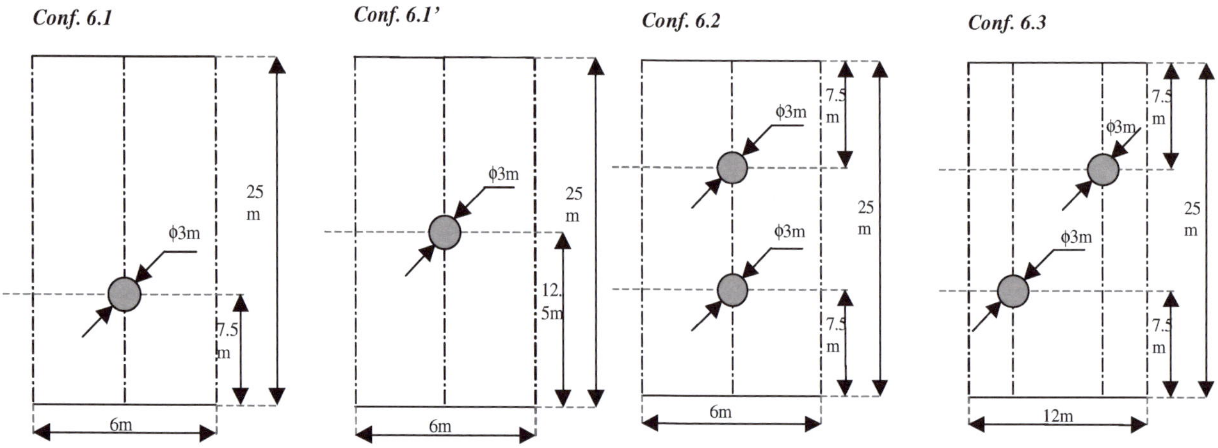

CONF.6.1	Toffice	Toutlet Toffice	Values at 1.5m height			
			ΔT (horizontal)	v (max.)	v (av.)	ΔT (vertical)
up/up	24.9°C	-0.5°C	2°C	0.51m/s	0.11m/s	0.2°C
up/down	25°C	-0.5°C	2.3°C	0.51m/s	0.12m/s	0.2°C

CONF.6.1'	Toffice	Toutlet Toffice	Values at 1.5m height			
			ΔT (horizontal)	v (max.)	v (av.)	ΔT (vertical)
up/up	25.3°C	0.3°C	1.3°C	0.5m/s	0.12m/s	0°C
up/down	25.5°C	0.2°C	1.3°C	0.5m/s	0.12m/s	0°C
down/up	25.6°C	0°C	1.5°C	0.16m/s	0.06m/s	0.6°C

CONF.6.2	Toffice	Toutlet Toffice	Values at 1.5m height			
			ΔT (horizontal)	v (max.)	v (av.)	ΔT (vertical)
up/up	26.2°C	-1.1°C	1.1°C	0.37m/s	0.11m/s	0.2°C
up/down	25.5°C	-0.4°C	1.1°C	0.39m/s	0.12m/s	0.1°C
down/up	24.9°C	0.1°C	1.9°C	0.17m/s	0.06m/s	0.7°C
down/down	28.1°C	-2.1°C	2.2°C	0.13m/s	0.04m/s	2°C

CONF.6.3	Toffice	Toutlet Toffice	Values at 1.5m height			
			ΔT (horizontal)	v (max.)	v (av.)	ΔT (vertical)
up/up	25.9°C	-0.7°C	1.6°C	0.44m/s	0.12m/s	0°C
up/down	25.7°C	-0.5°C	1.5°C	0.44m/s	0.12m/s	0.1°C
down/up	26.8°C	-0.2°C	2.2°C	0.23m/s	0.1m/s	0.7°C
down/down	27.5°C	-2.3°C	3.1°C	0.19m/s	0.08m/s	0.6°C

Where:

ΔT (horizontal) is the maximum difference of temperature on the plane at z=1.5 m, this is obtained as the subtraction of the maximum and minimum temperature of any point on that plane, and it shows how large the horizontal gradient is.

v (max.) is the maximum velocity on the plane at z=1.5 m, this is an important and limiting parameter as the velocity must be kept under a maximum value, above which human comfort is not good.

v (av.) is the average velocity on the plane z=1.5 m.

ΔT (vertical) is the vertical gradient of temperature at z=1.5 m, this is useful to measure the temperatures stratification

---

Best solutions, in terms of the four parameters above, were obtained on Configuration 6.1´ and Configuration 6.2. Best values for short-circuit control were given by Configuration 6.1´, since an adequate distance between inlet and outlet made the air get warm while it left the room. Thus Configuration 6.1´ kept good values in all the solution parameters of the problem, being probably the best of the possibilities. Configuration 6.2 could be a good option too. It kept smaller horizontal temperature gradients as well as smaller velocities. This could be caused by the utilisation of a couple of towers, which helped to distribute the cool air much more efficiently and with less mass flux per tower. Main inconvenience came as a consequence of towers situated closer than before to the walls. Thus short-circuit risk became more likely as seen on Toutlet-Toffice negative values.

Control system

The PHDC system only operates when the supply air temperature is, at least two degrees below the indoor temperature. In such cases, when the supply relative humidity is greater than a prescribed value, the control modulates the number of micronisers so that this maximum is not exceeded.

A second modulation of the number of micronisers into operation appears when the outdoor air temperatures (and consequently the load) are lower than the design value.

Even if the load is null, the PHDC system could operate so as to provide the ventilation needs of the building, in this way the electric consumption of the fans is saved.

Also, in some locations it could be possible to use the evaporative effect together with a good control system so as to provide a night cooling effect that reduces the cooling energy demand during the first hours in the morning. In this case the design has to be precise to avoid unfavorable effects like flow reversal at night. The night cooling effect can be evaluated using together the PHDC air flow and energy and comfort software.

The design temperature for cooling requirements mentioned below could be the peak temperature in summer for the last 10 years; this is a good value to have the maximum cooling load with very low percentage of having a day worst than this, the drawback is that this design temperature leads to oversized systems. In order to avoid this problem of oversizing, other criteria for the design temperature are the mean of the maximums or the daily mean during the summer, this leads to a smaller system but more probability of occurrence of a day worse than the design one. In any case, all the cases mentioned could be evaluated using together PHDC air flow and PHDC energy and comfort.

6.2 CLIMATIC QUALIFICATION APPLICABILITY

A first estimate of the applicability of the PHDC techniques in a certain location can be obtained from the climatic data (section 1.1). There are many variables that can be used as indices to characterize the cooling potential[2]. The cooling energy that can be transferred from the air -working fluid- to the building over a certain period of time is given by:

$$Q = \int_t \dot{m}(t) c_p \left(T_{indoor}(t) - T_{outlet}(t)\right) dt$$

Where:

$\dot{m}(t)$: Air mass flow rate depending on time, (kg/s).

c_p : Specific heat of the air at constant pressure, (kJ/kg °C).

$T_{indoor}(t)$: Indoor temperature depending on time, (°C).

$T_{outlet}(t)$: Outlet temperature depending on time, (°C).

If the mass flow rate and the indoor air temperature can be assumed constant during the period, the corresponding specific cooling energy in degree-hours (DH) can be written as:

$$\frac{Q}{\dot{m} c_p} = \int_t \left(T_{indoor} - T_{outlet}(t)\right) dt$$

Obviously, the minimum value of the outlet temperature would be (in a theoretical optimum situation) equal to the temperature of the environmental sink. Consequently, an upper boundary of the specific cooling energy would be:

$$\left(\frac{Q}{m c_p}\right)_{máx} = \int_t \left(T_{indoor} - T_{sink}(t)\right) dt$$

The next figure shows the difference between the outdoor dry and outdoor wet bulb temperature in degree-days in Spain. (fig. 135)

Potential of Evaporative Cooling to Reduce the Temperature of the Air.

The red area represents the places where the use of evaporative systems to cool the air is more applicable.

The next map shows the potential of cooling the building with a tower that theoretically could obtain the air in wet bulb conditions. (fig. 136)

Potential of Evaporative Cooling to Reduce Cooling Demand.

The red area represents the places where the reduction in cooling demand would be higher using evaporative cooling.

The map (fig. 137) shows the values of the maximum specific cooling energy obtained from July to September, assuming an indoor temperature of 25°C. The integral is extended to the operation period (from 8a.m. to 6p.m.) of an office building used as reference.

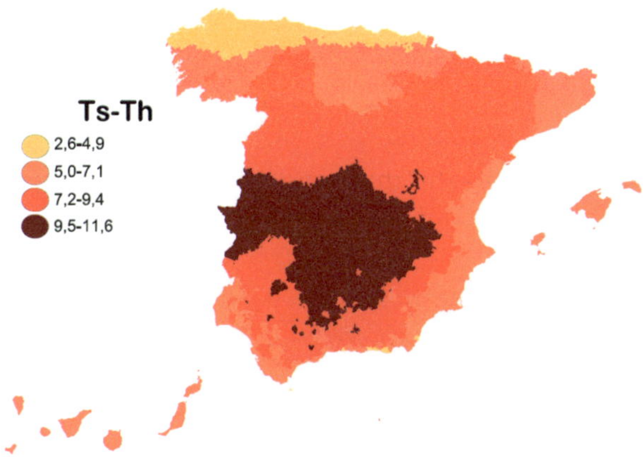

135

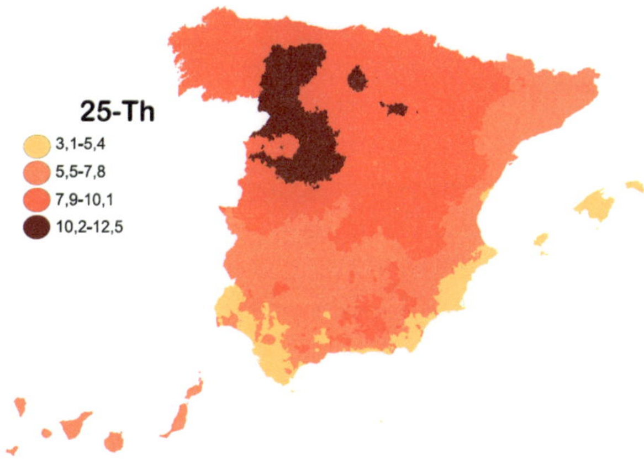

136

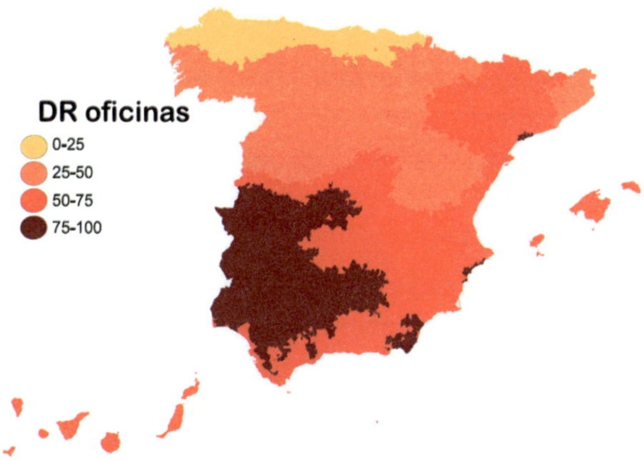

137

135. Map of Spain showing the difference between outdoor Dry Bulb Temperature vs. outdoor Wet Bulb Temperature. Average over 24hours for June – September.

136. Map of Spain showing the difference between indoor Dry Bulb Temperature (taken as 25°C) vs. outdoor Wet Bulb Temperature. Average over 24hours for June – September.

137. Map of Spain to show Cooling Energy Demand in Office Buildings in kWh/m2. Internal gains of 15 W/m2 during 12 hours a day.

PHDC Air Flow: User Manual

The image shows the simplified schematic section used by the software. There are three parts: the inlet on the left, the central area, and the outlet on the right. By clicking on there different sections, a properties window will be activated. Here, changes can be applied in an easy and straightforward way.

By clicking the Properties tab, indoor and outdoor conditions can be modified, such as wind velocity, location's altitude above sea level, and the height of every floor.

By ticking the boxes, the inlet and/or outlet towers can be activated. An image of the corresponding tower will open, where properties can be modified as it will be described in the next page.

Number of floors. Up to 4 levels.

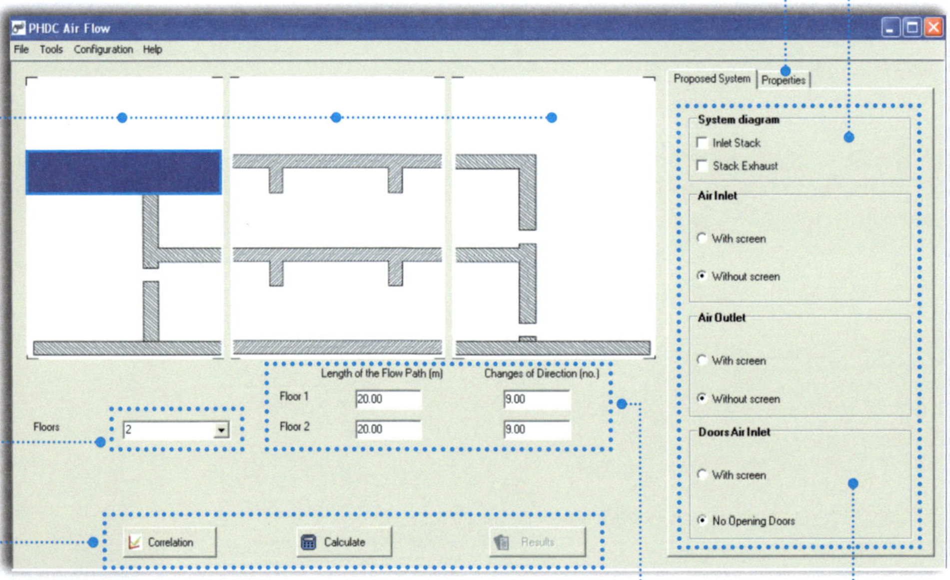

These buttons will analyze the data. "Calculate" will generate results and show them in a new widow, which can be recalled by clicking "Results". The "correlation" button will obtain a correlation for each storey useful for the PHDC energy & comfort software.

The distance between inlet and outlet (total length of air path), can be defined here, as well as the number of changes in direction.

In this section, the type of opening for inlet, doors inlet, and outlet can be configured: with or without grilles and with or without doors.

Inlet tower: When it is activated, this window below appears.

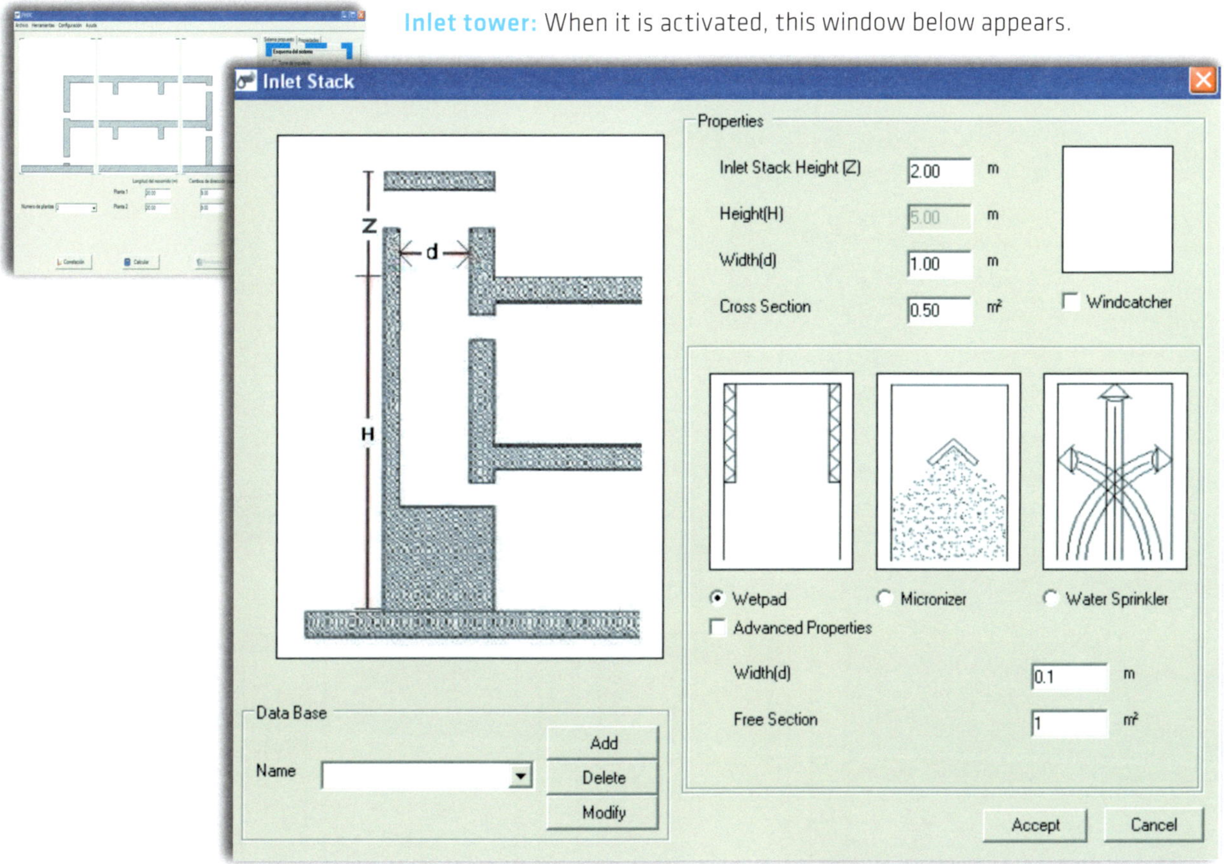

The "Properties" section, on the top right, shows the geometric properties of the tower. The "Height"(H) is the total height, including all levels, and excluding the top of the tower. The chimney height, which stands out from the top of the building, is defined in "Inlet Stack Height" (Z). The width of the tower (d), and the area of the effective downdraught (d) can also be configured. Additionally, in this window the type of air catcher can be defined, if needed, by ticking the "Windcatcher" box. The software offers five different types of catchers, and a sixth possibility which is configured by a user defined pressure coefficient. See section 2.1.3 to understand the models and characterization of each catcher.

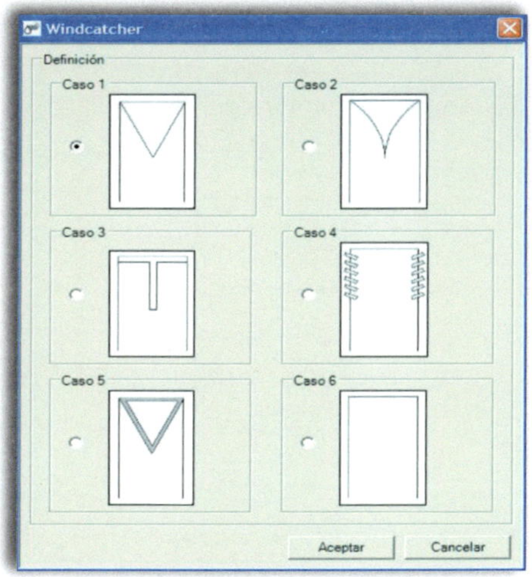

Evaporation system

The evaporation system needs to be chosen: *micronizer, water sprinkler o wetpad.*

Each evaporation system has an associated window where the user can configure the properties.

For the micronizers required properties are: height at which micronizers are located (p), droplets diameter (µm), and water flow (kg/sec).

"Diameter of droplets " refers to Sauter's mean diameter, that is to say, to make alike the drops to spheres with a such diameter, which remains the relation volume/surface of the drops.

Besides to the "diameter of droplet" there is a checking box, clinking here the user can alternate between this data and the "distance of 50% of water evaporation". This parameter "Distance" together with the "diameter of droplet" before commented, define the dependence of the technology used in the process of micronizer. Definitively, "Distance of evaporation of fifty per cent of the impelled wealth " is a measure of the efficiency of this device, where it are considered to be the rest of parameters of the same one.

By checking the box for "distace of 50% of water evaporation", the user can alternate between this option, and the "droplet diameter" option. It is recommended to read point 2.3.1 in order to understand the differences between these cases.

THE EVAPORATION SYSTEM NEEDS TO BE CHOSEN: MICRONISER WATER SPRINKLER OR WETPAD.

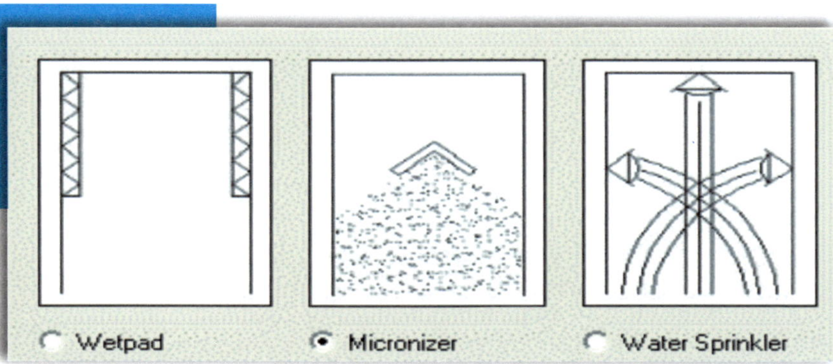

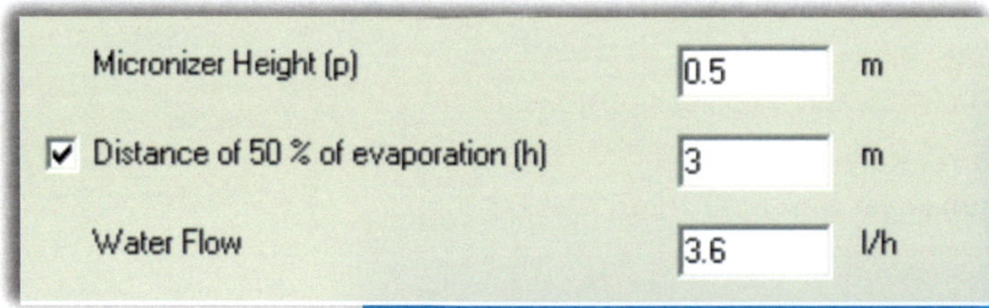

BY CHECKING THE BOX FOR "DISTANCE OF 50% OF WATER EVAPORATION", THE USER CAN ALTERNATE BETWEEN THIS OPTION, AND THE "DROPLET DIAMETER" OPTION.

The wetpads properties require the thickness and the free section, as explained in 2.3.3. Without any further data, the user could simulate a default wetpad. By clicking on the "Advanced Properties" the user can modify the default parameters (see parameters of various commercial wetpads in 2.3.3.

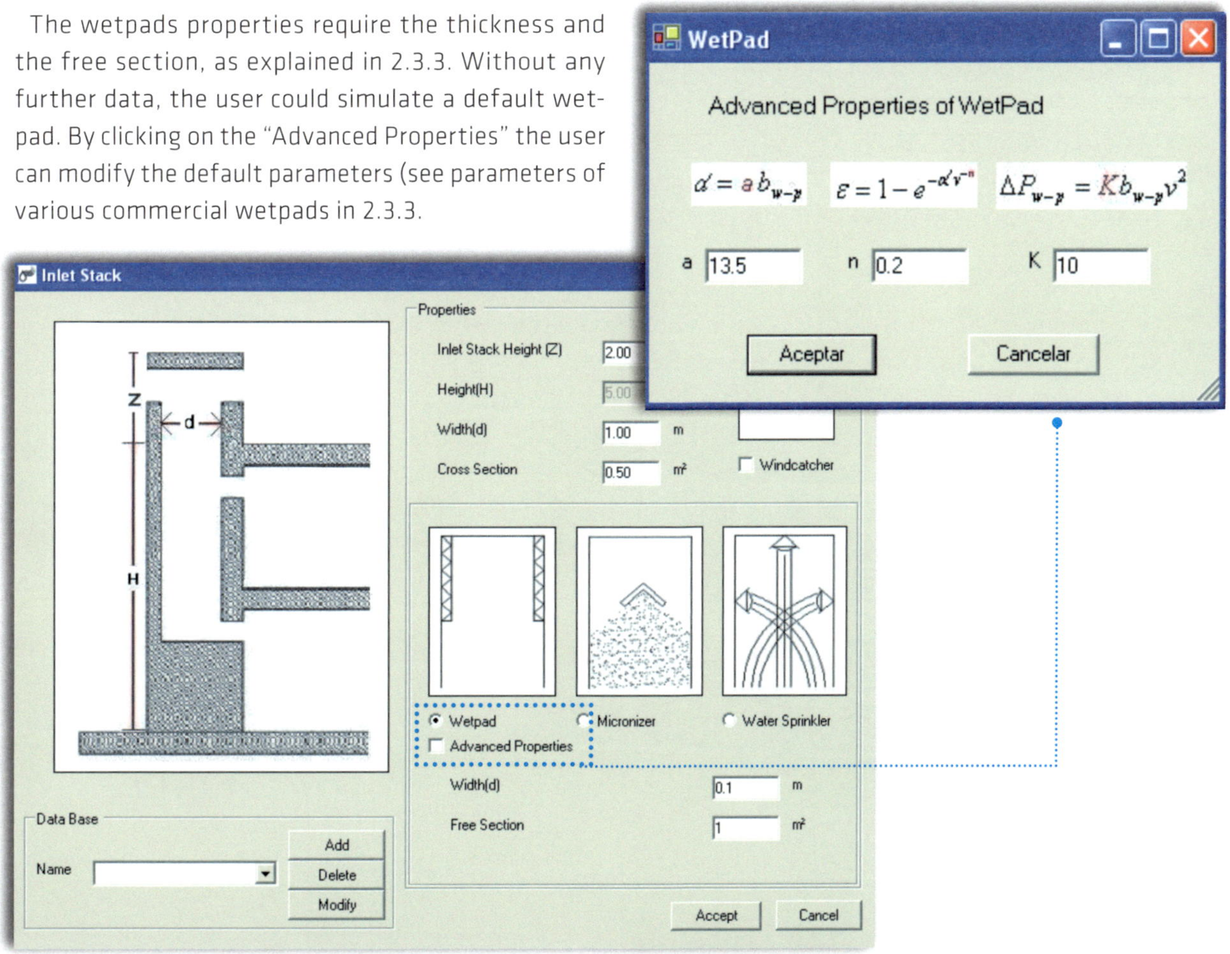

The sprinkler properties require the height at which sprinkles are located (p), droplets diameter (μm), and water flow (kg/sec). How these parameters are used to define the sprinklers is described in section 2.3.2.

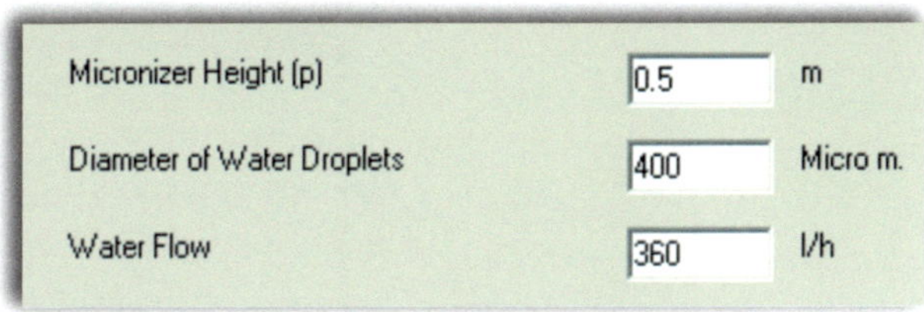

In each case the properties are schematically drawn in the window besides the properties square. It is possible to save and/or load a case for using this tower configuration in a future, this option is below the graphic.

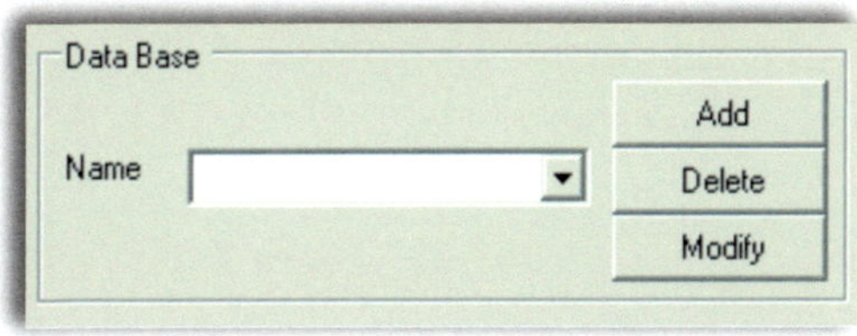

Exhaust tower: When it is activated, this window below appears.

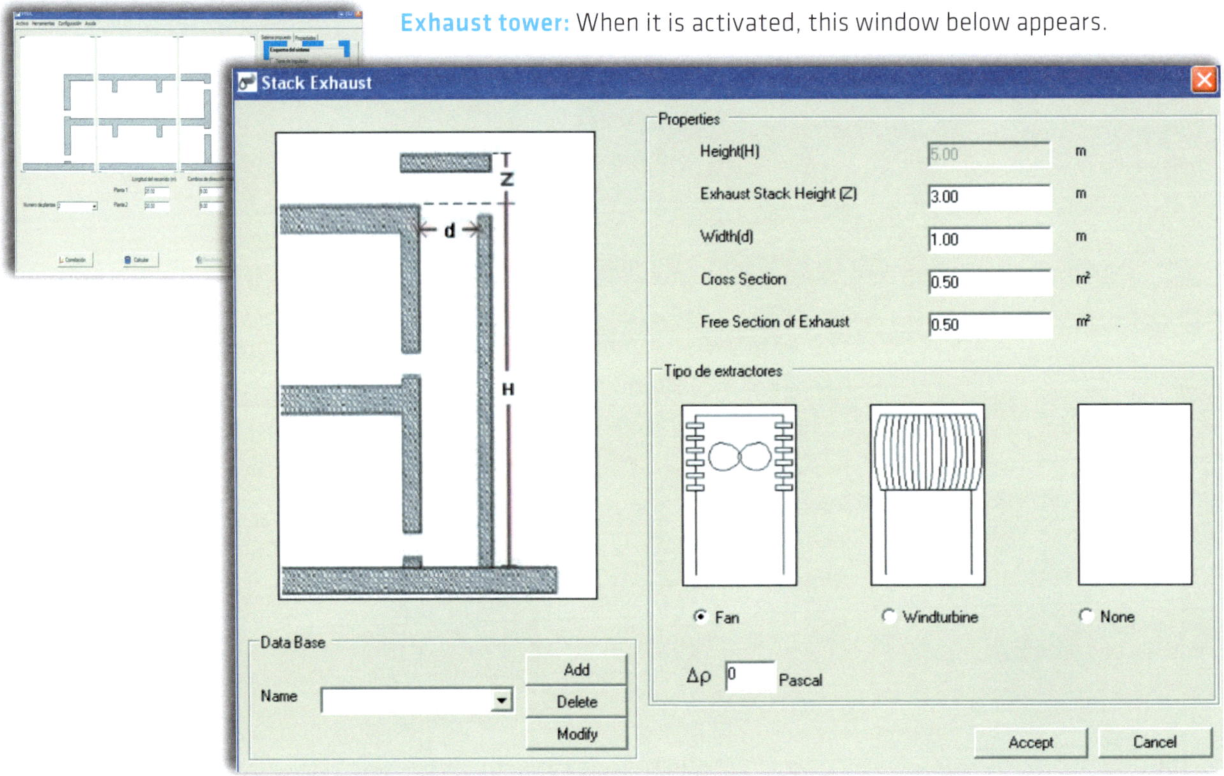

This screen is very similar to that of the "Inlet Tower". Here, the user can configure the exhaust tower properties. The "Height"(H) is the total height, including all levels, and excluding the top of the tower. The chimney height, which stands out from the top of the building, is defined in "Exhaust Stack Height" (Z). The width of the tower (d), and the area of the tower section (d) can also be configured. A "Free section of exhaust" area need to be defines, to characterize the pressure losses, when the air goes through the wet pad.

Types of Extractors.

There are three offered possibilities: mechanical extractor (fan), wind turbine, or none. The first two are defined by their extraction power (measured in Pascals).

Openings:

All the inlets, outlets or passing elements (like doors) are characterized using their coefficient of flow (c [m3/s·Pan]) and their exponent of the flow (n). (See 2.4). These values can be different from one storey to another for the user to design the openings in order to obtain a homogeneous flow in all the storeys.

The program allows to change the parameters of the grids of entry and exit of the air into every floor, on having activated the existence of grids, the program will present an appearance as the showed one.

In the same one it is possible to observe a drop-down menu in which the floor will be chosen, once chosen, in the top picture we will give a value and will certify pulsating on the icon with the symbol of the definite parameter. The sequence will have to repeat for every floor and parameter, if values will not be assigned by fault.

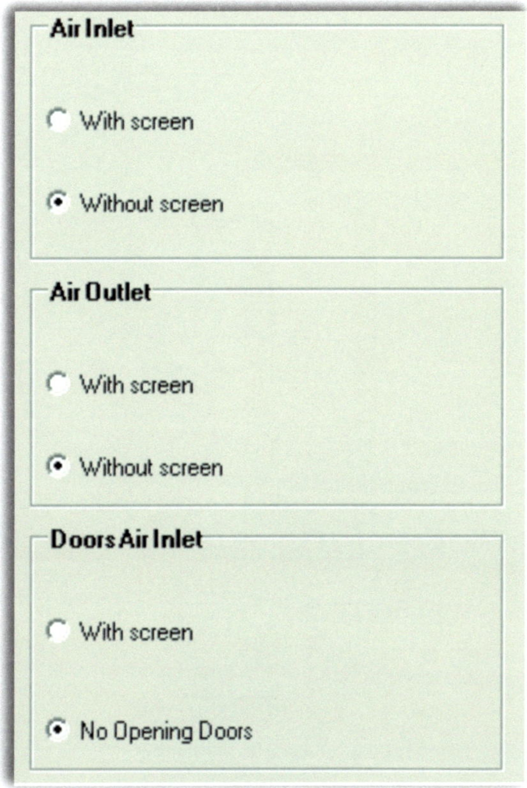

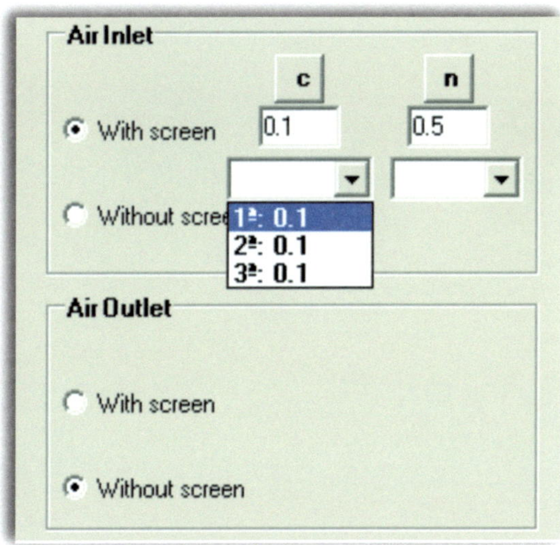

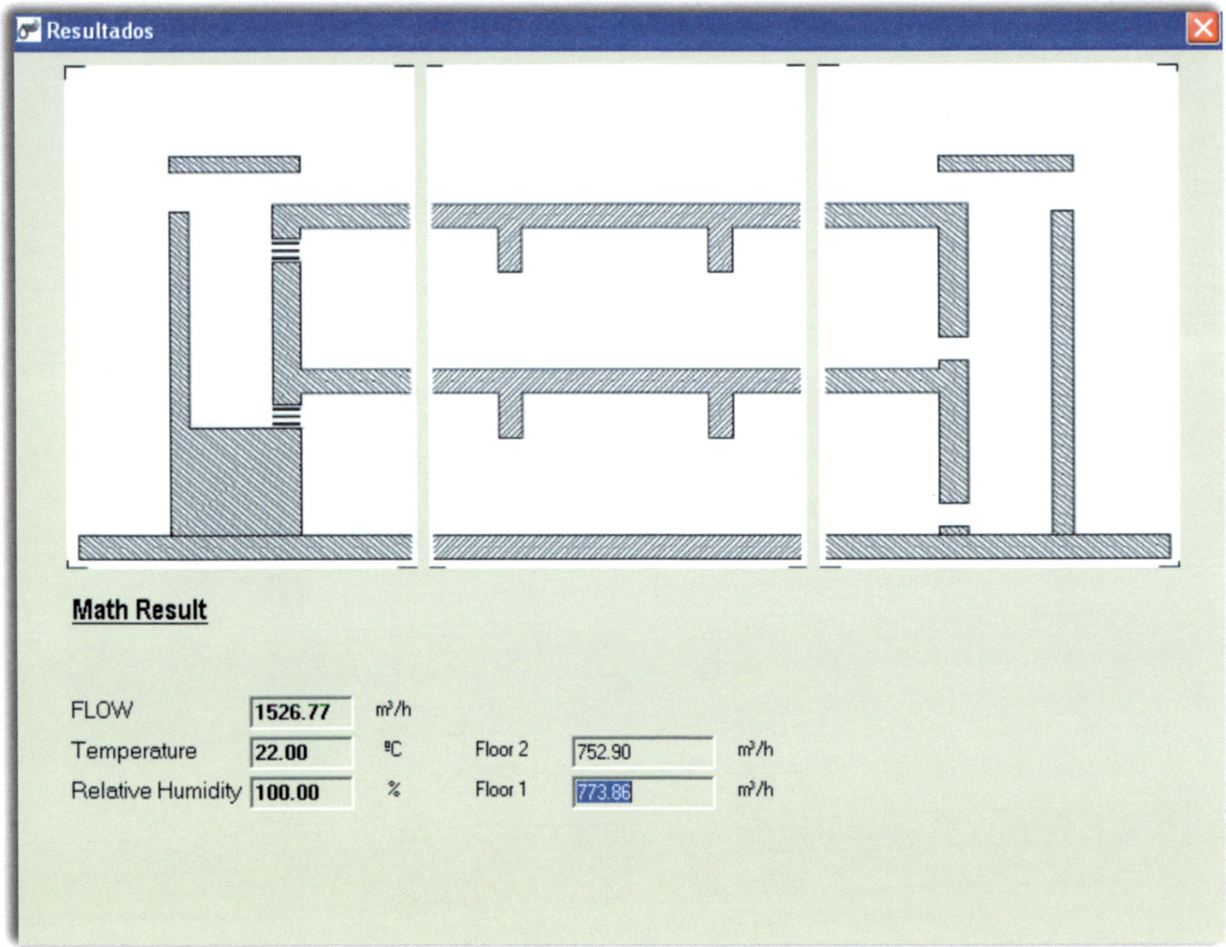

Calculations and results: *Calculations and results: the screen above opens after calculations have been requested.* From left to right it shows the results of the inlet tower (total flow, inlet temperature and relative humidity), and the air flow in each storey.

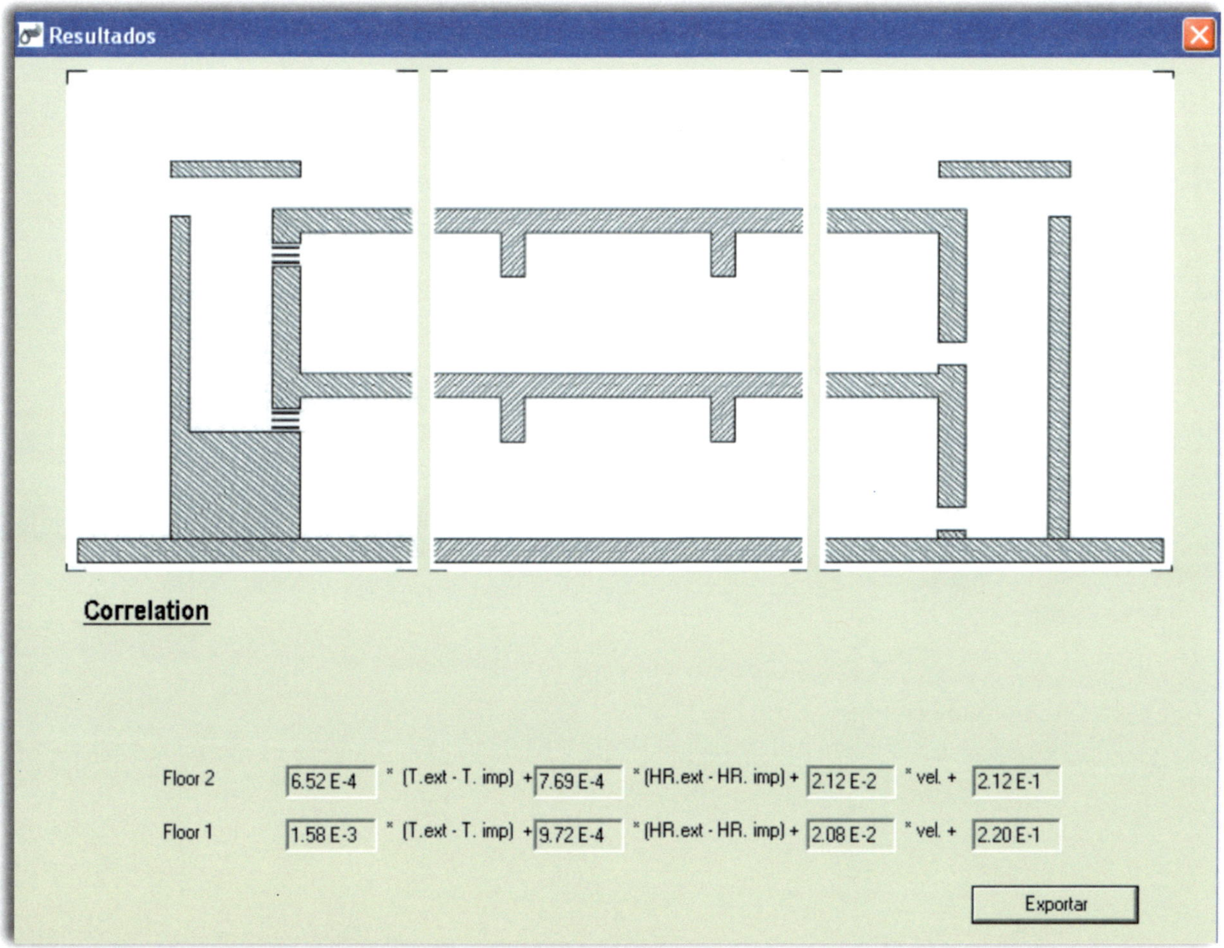

Correlation: the screen above opens when clicking on the "Correlations" button. Here, a correlation for obtaining the airflow in each office is shown. This correlation is used in the PHDC energy and comfort software by clicking in the "Export" button.

CHAPTER 7
COMPONENT DESIGN AND CONTROLS

7.1 CONTROL AND INTEGRATION

EFFECTIVE INTEGRATION REQUIRES CLOSE COLLABORATION BETWEEN ARCHITECTS AND ENGINEERS

Having developed a clear overall downdraught cooling strategy as early as possible in the design process, careful consideration needs to be given to component design and integration, as part of the overall design development. Achieving successful downdraught cooling in practice depends on the detailed design and specification of the different components and on the careful integration of the ventilators, actuators and controls in the building.

Effective Integration

OCCUPANTS WILL BE MUCH MORE TOLERANT OF CONDITIONS IN A BUILDING WHERE THEY PERCEIVE THAT THEY HAVE SOME INFLUENCE ON THEIR SURROUNDINGS

Effective integration requires close collaboration between architect and engineer. For example it is important to agree who specifies what. An architect will often specify the window, and an engineer the actuator and control. Will they be compatible? How will they be put together? A factor often overlooked is that ventilators and actuators are joined by linkages. Experience suggests that in many installations, linkages and fixings are major points of weakness and sometimes they fail entirely.

When architect and engineer have agreed on the performance required of the windows and/or motorised dampers, there are many advantages if the window/louvre supplier is given responsibility for integration of vent, actuator and control. Where they are not, vent and actuator suppliers all recommend early dialogue, during which issues can be raised, for example providing suitable fixings for the actuator, suitable linkages, and the need for any reinforcement of the frames. The individual components are rarely designed to be part of an integrated system, for example: windows, ventilators and rooflights which initially may have been designed for manual control need to be matched with a suitable actuator to operate successfully in response to changing ambient conditions. There are fewer problems perhaps when the opening, actuator and controls are specified as a fully-integrated system, so that installation can be largely maintenance free.

With water misting systems, it is important that the effect of the 'cone angle' of the spray nozzle or microniser is assessed in order to avoid wetting of adjacent building components or structure, and subsequent dripping into areas below. The cone angle from the nozzles also needs to be set to avoid adjacent drip trays.

Where possible, it is desirable to combine automatically controlled ventilators (for example at high level or out of reach of occupants) with occupant controlled ventilators. Occupants will be much more tolerant of conditions in a building where they perceive that they have some influence on their surroundings[1]. Equally, allowing manual override of automatically controlled vents can help, and still provide effective control (with control going back to a default position after a set period).

Low and High Pressure Misting Systems

Misting nozzles which rely on hydraulic pressure alone require a high pressure (25 – 40 bar or more) to create droplets of 30 microns or less. Hydraulic nozzles tend to be made of either brass or stainless steel, requiring appropriate pumps and pipework with compression fittings. Such a system can be relatively expensive to install, but can then run successfully for many years, requiring relatively little maintenance.

The alternative of using pneumatic technology (compressed air blown through nozzles supplied with a low pressure water supply) can use relatively low cost plastic nozzles to create droplets of less than 15 microns in diameter. Pneumatic technology of course requires compressors (instead of pumps), but does have a number of advantages over hydraulic systems, including:

- relatively low cost system components (namely pumps, pipes, connections and nozzles) when compared to those in the high pressure systems.
- generally higher robustness of the system as a whole. On the one hand, the lower pressure puts the system into much less stressful conditions; on the other hand, the much wider orifice reduces the risk of clogging.
- no final jet "spitting". High pressure systems tend to "spit" a final short water jet when they are shut down as a result of the decrease in pressure. This issue makes the system unfit to many applications where these droplets may fall from above onto people or equipment below.

Ventilation openings

Having defined the cooling strategy, the choice of vent opening type can be made. The strategy is likely to have identified the need for different airflow rates in summer and winter. Decisions will also have been made about which ventilators are manually or automatically controlled. People like to be able to make alterations to their environment, so providing some manual control at low level is often a good idea. Conversely, high level vents are difficult to reach and their status (open/closed) is sometimes difficult to determine, so automatic control is advisable. This arrangement also means that night ventilation can be controlled automatically and securely. Windows and rooflights can be made to shut tight relatively easily, and in comparison with louvres and dampers they have a shorter crack length, and more effective seals. Window stays vary enormously in terms of their 'throw', geometry and robustness. Care must be taken to ensure that the effective free area will be achieved in practice. (fig. 138) Most patterns of window were originally designed for manual operation. Windows used for automatic control may require adaptation to accept motorised actuators, and strengthening to accommodate the forces applied. Detailed guidance on window and ventilator design is provided by CIBSE in *"Natural Ventilation in Non-Domestic Buildings"*[2].

7.1.1 VENTILATOR DESIGN

138. Ventilator design definition of 'effective opening'.

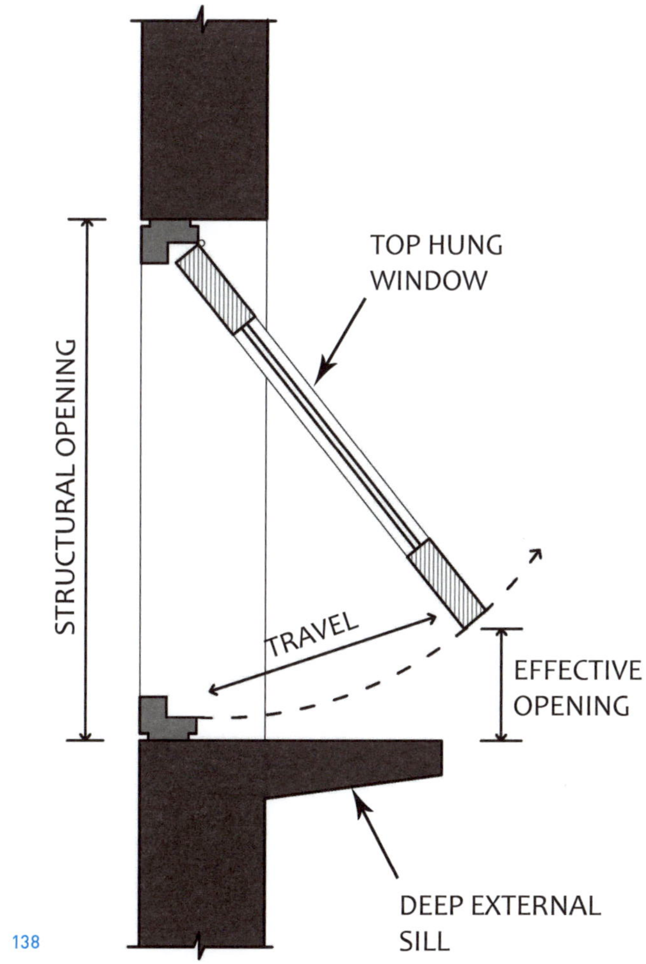

Different window and louvre designs may be assessed under the following criteria.

Ventilation Capacity. This is related to the way the ventilator opens and the surrounding head, cill and jamb details. Determining the effective area for a particular window type needs careful consideration. With top hung vents for example, the triangular opening each side of the open window is significant. In the event, an estimate is usually the basis for calculations. Figure 141 illustrates the difference between structural opening, throw, and effective opening.

Controllability. Where it is considered desirable and acceptable for occupants to control opening of windows, dampers or louvres, stays should be both adjustable and robust. With manual and automatic operation, the stay increases the angle of opening of the vent and the effective area will increase, but the relationship is not linear, and this will also vary according to the opening type[3].

Comfort. Generally, the use of opening vents results in high levels of user satisfaction, when local occupant control is part of the strategy. However, the use of the same opening for winter and summer ventilation may be unsatisfactory in certain situations. This can be a reason for maintaining low level openings under manual operation, and high level openings under automatic control. The provision of automatically controlled high level window openings for night ventilation can incorporate manual override to provide occupants with a greater degree of daytime control.

Security. It could be argued that open windows are only likely to be a security problem at ground or first floors of a building. Restricting the length of throw of stays or actuator arms may be sufficient in many situations.

Sealing. The sealing of windows is usually achieved by EPDM expanded rubber gaskets. In aluminium framed windows the tolerance achieved is usually very fine and good sealing is possible. However, with some steel framed and timber windows, racking or twisting of the frame can occur, particularly with large windows, resulting in poor sealing. Motorised dampers and louvers do not shut as tight as most windows, owing to: longer crack lengths, difficulties with rotating seals, and problems with mechanical strength and closing forces. This may lead to excessive uncontrolled infiltration, and increased cooling (and heating) loads.

Integration with Vent Actuators. The choice of window type and its integration with different actuator options and (where applicable) internal blinds, requires careful consideration if the performance of one or other of these components is not to be compromised. Studies of a number of different installations[4] have revealed that a good general rule is to have one actuator per ventilator. If one actuator powers several ventilators, then difficulties in obtaining good initial alignment - together with accumulating backlash in the linkages - nearly always means that not all the ventilators close tightly, leading to problems with air tightness. This can be aggravated by racking of the ventilator frames, by distortions from poor fixing, by flexing of the linkages, lever arms and brackets, or from shock loads from wind or at the ends of travel.

THE RESISTANCE TO AIRFLOW OF BAFFLES, LOUVRES AND INSECT/BIRD MESH MUST BE ALLOWED FOR IN THE CALCULATION OF EFFECTIVE FREE AREA

7.1.2 SHAFTS, DUCTS AND COOL TOWER DESIGN

Downdraught cooling strategies require the introduction of fresh supply air at high level, normally 2-3m above the ceiling of the highest served floor. (figs. 139 - 140) The termination of shafts or cool towers will normally be well above the roof level, in a zone of negative pressure (suction) in windy conditions. However, positive pressure at the supply entry point is desirable to promote a daytime downdraught, and this has led to attempts to 'catch' the wind[5]. While there are numerous historic examples of wind-catchers in the extreme climates of the Middle East, catching the wind can also lead to over-pressure and uneven air distribution.

Supply Air Termination

Misting nozzles require a height of 2-4m for water droplets to be completely evaporated, depending on the ambient conditions. Therefore, the supply air termination must stand at least 3m above the roof level of the top floor of the building, if the top floor is to be served by the PDEC system. It has already been established that the temperature difference between the supply air (the tower exit temp) and the room will be small (possibly only 2-3°C). In order to achieve substantial cooling, the openings into the shaft therefore need to be large to achieve the required air volume flow rate. However, measures to baffle the wind, and to prevent the ingress of rain and possibly dust, must not be allowed to reduce the effective free area of inlet opening. The resistance to airflow of baffles, louvres and insect/bird mesh must be allowed for in the calculation of effective free area. Of course, where perimeter towers and linked plena are used, then restrictions on dust and rain ingress may not be so strict, as they can be separately cleaned. (fig. 141)

139

140

141

139. Lake View Library PDEC tower.

140. Exhaust Air termination with motorized glazed ventilators, Queens Building, De Montfort University, Leicester UK.

141. Perimeter misting towers with fixed louvers serving the auditorium at the CII Centre, Bangalore, India.

Wind Baffles

An assessment of the micro-climate of the site will confirm whether the wind and associated driving rain or dust penetration are likely to be a significant factors when developing the detailed design of the downdraught cooling supply towers. Much will depend on the location and the building type (for example, a low rise building within an urban area is unlikely to require wind baffles). However, when wind baffles are required to deflect the wind and prevent it from counteracting the downdraft effect, it is likely that overlapping layers will be required to be effective. Experience suggests that louvres on their own will be insufficient. The baffles specified for the Torrent Research Centre, Ahmedabad, were vertical half-round ceramic pipes, which help to turn the wind back on itself to further reduce the localized wind velocity. The half-round pipes were located outside fixed louvers and an inner layer of insect mesh, as shown in figure 142 below.

AN ASSESSMENT OF THE MICRO-CLIMATE OF THE SITE WILL CONFIRM WHETHER THE WIND AND ASSOCIATED DRIVING RAIN OR DUST PENETRATION, ARE LIKELY TO BE A SIGNIFICANT FACTORS WHEN DEVELOPING THE DETAILED DESIGN OF THE DOWNDRAUGHT COOLING SUPPLY TOWERS

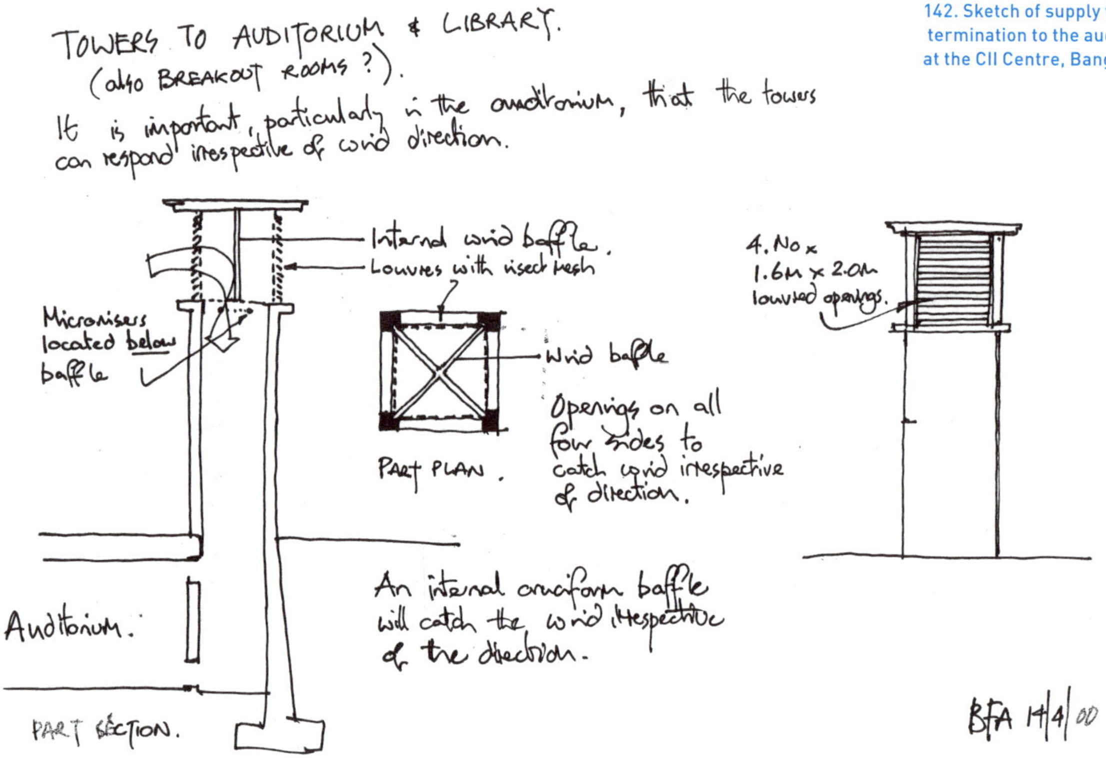

142

142. Sketch of supply tower termination to the auditorium at the CII Centre, Bangalore, India.

Wind Catchers

In locations where wind is reliable and thought to be advantageous to increasing airflow through the building, a wind catcher can be incorporated at the top of the supply tower. Traditionally, wind catchers have been of two types: towers designed to catch the wind irrespective of wind direction (used in parts of Middle East where the direction of wind is variable); (fig. 145) and towers designed to catch the wind from the direction of a reliable prevailing wind (used in parts of Pakistan).

Omni-directional wind-catchers

Are found In traditional Iranian wind towers, where a cruciform baffle is invariably located at the top of the tower. The baffle not only directs the wind downwards, but also causes any dust in the wind to fall, rather than be carried further into the building. The internal baffle also allows for some air to be simultaneously exhausted from the leeward side of the tower. Wind catchers are considered in the PHDC software as described in chapter 6 and annex C.

The benefit of a wind-catcher is that it can significantly increase the volume flow rate (and hence cooling potential) of the tower. Studies at the Blaustein Institute, Ben Gurion University, Israel[6], have explored the implications of wind-catcher geometry on the volume flow rate. (fig. 143)

Further studies of wind-catcher geometry were undertaken by MCA as part of the PDEC research project[7], which explored the use of a fabric 'bell' shaped 'straightener' within the tower shaft. (fig. 144)

Where provided, the internal wind-catching baffle must be located above the microniser nozzles, otherwise the baffle will become wet and cause dripping down the tower. This could be a significant dis-incentive, as it effectively increases the required height of the supply tower from 3 to 6 metres above the top floor of the building.

THE BENEFIT OF A WIND-CATCHER IS THAT IT CAN SIGNIFICANTLY INCREASE THE VOLUME FLOW RATE (AND HENCE COOLING POTENTIAL) OF THE TOWER

Mono-directional wind-catchers

Where a very reliable prevailing wind exists, then mono-directional wind-catchers can be effective. Historically, such towers have been built in the Sind province of Pakistan and along the coast of Oman.

143. Wind catcher geometrical variations.

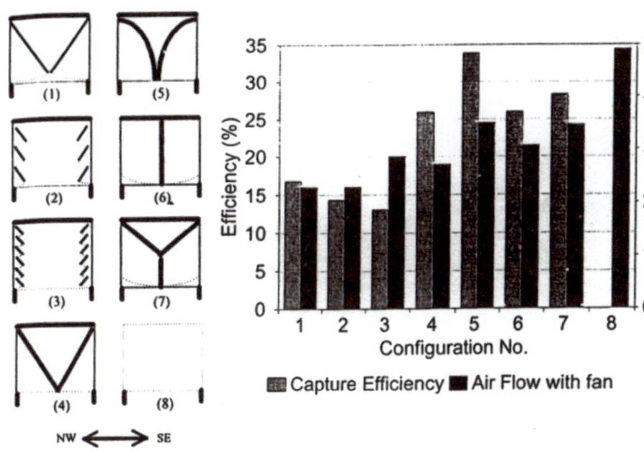

143

144. Experimental windcatcher.
145. Windcatcher tower in Dubai.

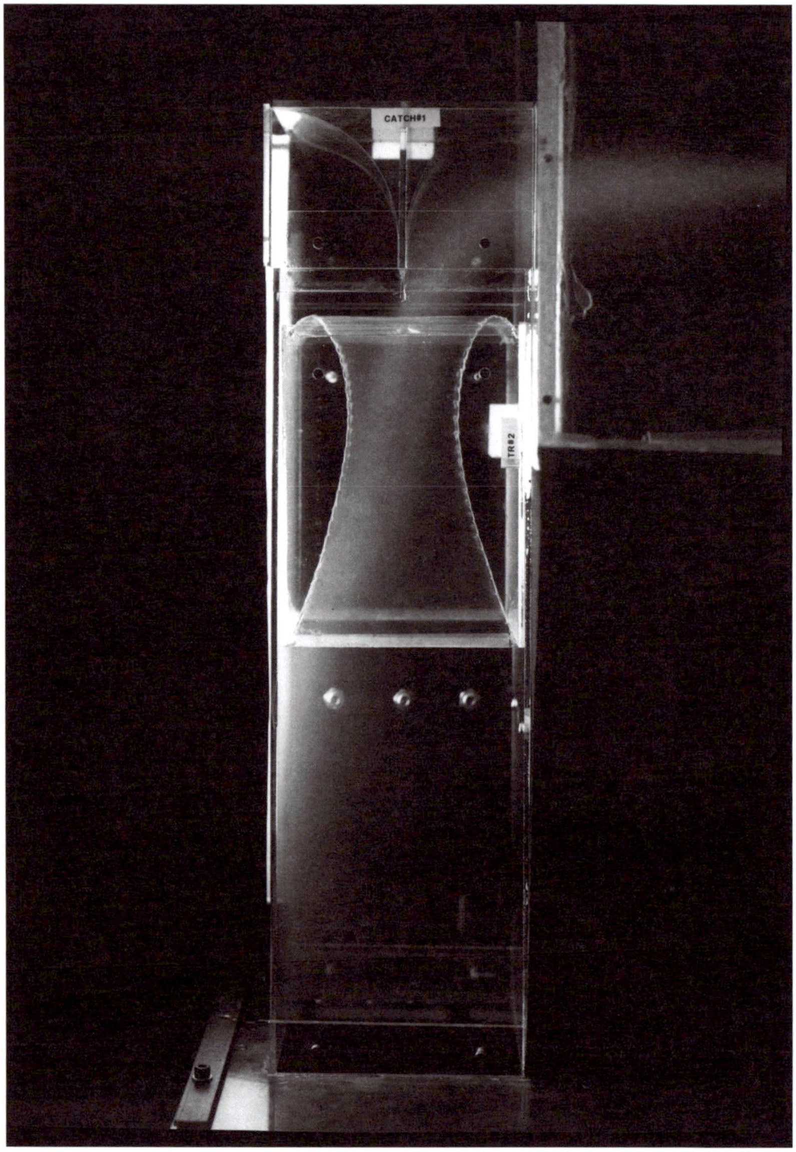

144

145

7.1.3. CONTROLS

Control systems, either manual or automatic, should be kept as simple as possible, consistent with the successful operation of the building. Training is required for the owner's representative or facilities manager. Given adequate time for commissioning and training, the operation of most PHDC buildings has been reliable. The amount of air flowing through a given opening in a given position can also vary widely, as the driving forces (wind, natural buoyancy) change with indoor and outdoor conditions. In addition, the internal resistance to cross-ventilation will vary as doors etc are opened and closed. With the wide variations in the amounts of air needed and the associated driving forces, it can be difficult to achieve the full range of requirements using the same ventilation device. For example, the large openings required for downdraught cooling in summer will need to be shut tight in winter with only minimum fresh air supplied. Also, while people may welcome air movement for its additional cooling effect in summer, they will complain of draughts in winter unless the incoming air is either well-diffused or preheated.

While high level openings may need to be automated, low level openings are preferably left in the control of the occupants.

Flow rates through ventilators

The flow rate through a ventilator is a function of the open area, but often this is not directly proportional to the actuator movement. In particular, the thickness of the ventilator and the way it fits into the building fabric can greatly affect the relationship. With a centre pivot or hinged window, the lower opening can be much restricted by the frame, the window cill and the reveal, particularly for the first 100 mm or so of travel. This can provide useful fine control, but it may restrict the required opening area for summer ventilation, particularly if safety considerations also restrict the permissible amount of window opening. For centre pivot or bottom-hung hopper windows, the upper opening may also be restricted by the reveal and by any internal blinds. External shading may also restrict flowrates and openings (it may also preheat the outside air, particularly if it is dark in colour and of a metallic finish).

Combined user and automated control

Occupants need to be informed about how the building is designed to work, so that the goals of comfort, air quality and energy efficiency are all achieved. Ideally, the strategy should be clear, straightforward and as far as possible intuitive. The control requirement will often change with the context for the user, for example, occupants sitting beside a window are most affected by the draughts from the lower casements, while those in the centre of the room are more likely to want to adjust the fanlights.

Suitable combinations of automated and manual control will not only provide backup in the event of BMS malfunctions, but will also empower the occupants, improving their adaptive opportunities, and making variations in conditions more acceptable than when there are no over-ride facilities. Window-type ventilators are often used at high level, but even where they are visible, it is not always easy to tell whether they are tight shut or a little bit open. In such situations, status lamps or other forms of indication are helpful. Dampers or louvres are frequently hidden from view: status feedback then becomes particularly important but is rarely included because of cost.

Maintenance. Lack of familiarity with the systems and the cost of seasonal maintenance appears to be the reason for some (previously successful) installations to fall into disuse. Building owners and facilities managers should be made aware that since these cooling facilities operate on a seasonal basis, they may require attention at the start and end of the cooling season, so that it is built into the maintenance schedule for the building as a whole.

OCCUPANTS NEED TO BE INFORMED ABOUT HOW THE BUILDING IS DESIGNED TO WORK, SO THAT THE GOALS OF COMFORT, AIR QUALITY AND ENERGY EFFICIENCY ARE ALL ACHIEVED

7.2 WATER SYSTEMS
Estimating water consumption in a PDEC installation

Like energy, water is a resource whose use must be managed carefully. This is especially true in the environments most suited to application of PDEC for cooling buildings – hot dry climates. The amount of water used for PDEC will vary substantially, depending on building use as well as climate. In addition, water consumption in a PDEC installation depends on:

- the volume of air supplied to the space being cooled,
- the thermal properties of this air, and
- the amount of water that must be evaporated to bring a unit volume of the air from its initial condition to the specified supply condition.

Analysis of the micro-climate (as described in section 1.1) will reveal the range of dry bulb temperature and humidity which is likely to prove acceptable within the building, within the context of adaptive thermal comfort theory. After internal and external heat gains have been estimated for a 'typical' summer day condition then the nominal cooling requirement can be estimated (as described in Chapter 5).

In a conventional HVAC system, the temperature of the supply air is quite low (16 - 20 oC) and is almost constant over time, irrespective of the external environmental conditions. As a result the volume of air required to remove heat gains from the space is quite low (typically 2-3 air changes per hour). In a direct evaporative cooling system (including PDEC), the temperature of the supply air is substantially higher, often only 1 -2°C cooler than the air in the space being cooled. As a result, the volume of fresh air required to remove heat gains to the building is significantly higher (typically 20 – 30 air changes per hour).

Consulting the psychrometric chart, we find that in our case, ambient air at 35°C and 20% relative humidity has an absolute humidity of only 7.3 g kg-1. If its humidity is increased by direct evaporation of water to 12 g kg-1 (the specified upper limit), it will have a temperature of approximately 23.3°C, which is 3.7 K less than the design temperature.

To calculate the amount of water required to cool the air from its initial (outdoor) condition to the supply state. This is simply the product of the difference in absolute humidity of the air in the two states, multiplied by the air volume flow rate:

(12.0-7.3)*40,950=192,460 grams, or about 192.5 litres per hour.

Finally, the above procedure must be repeated for the prevailing environmental conditions and changing cooling loads of the building for every hour of operation, and the resulting hourly values then added up to obtain the total water consumption over the period in question, such as a month or an entire summer.

Estimating the number of nozzles required

In using misting nozzles it is possible to evaporate all water within 2-3metres of the nozzles. This requires a droplet size of 30microns or less. Typical water consumption for a single nozzle is 5 litres/hour at 30microns, although this depends on nozzle characteristics and intended use. The AICIA PHDC 'Airflow' tool calculates the water consumption required under design conditions. For office buildings, this is between 1-10 litres/day per person in southern Europe. This can be compared with typical per capita consumption of 140 litres/day in Europe.

Water Treatment

The need for filtration and other water treatment measures obviously depends on local water quality, as well as the type of PHDC system envisaged. One of the benefits of using the 'shower tower' principle is that water quality is not an issue, although scale build-up is likely to occur in hard water areas.

Care must be taken to avoid micro-biological contamination, but as long as simple measures are taken this need not become a problem. Many engineers now refer to the Australian Standards AS/NZS 3666 Set: 2006.

Air-handling and water systems of buildings[8]

In order to mitigate the risk of microbiological contamination. Water generally should be taken from the mains supply, where the temperature is unlikely to be conducive to growth of Legionella or other organisms (which requires a temp of approx 32°C). Where storage is necessary, then water must be kept well below this temperature. Also, pipework to nozzles should drain down when not in use. A UV filter should also be installed on the line to the nozzles.

Maintenance of Nozzles

At TRC, where 316 stainless steel high pressure pipework and nozzles were installed, they soften and filter the water to reduce scale build up. The facilities manager at TRC (an engineer who has been with the project since its inception) has reported that blockage of nozzles at TRC has not been a problem – they are still using the same nozzles they installed 10 years ago. A monthly maintenance check includes removal and cleaning of any blocked nozzles – they have even developed a simple tool which enables them to unscrew nozzles from the walkways below, to allow cleaning and replacement.

146. PDEC water and air pipework schematic for Solar Decathlon Nottingham HOUSE for Madrid.

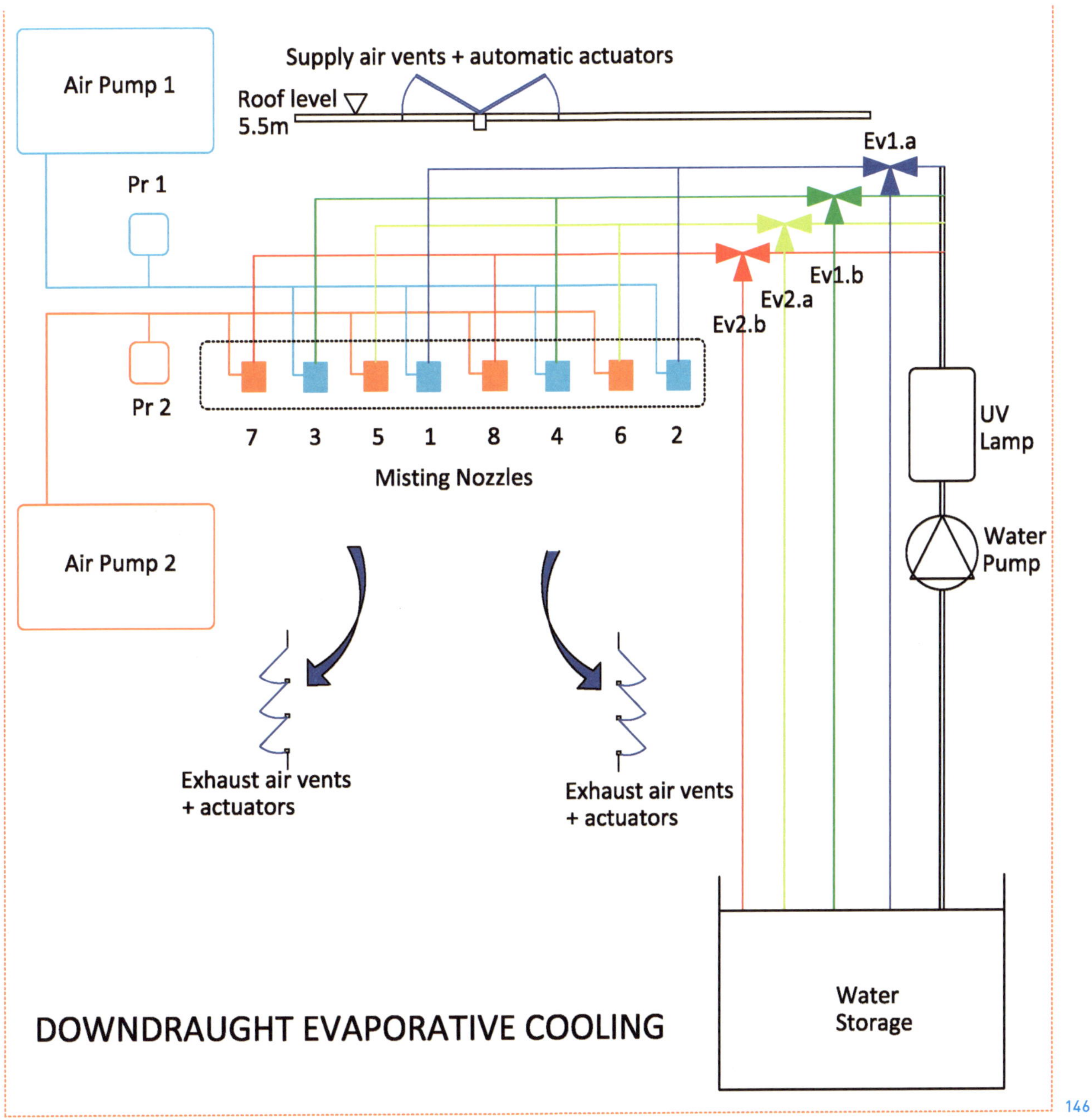
Air Pump 1
Supply air vents + automatic actuators
Roof level
5.5m
Ev1.a
Pr 1
Ev1.b
Ev2.a
Ev2.b
Pr 2
7
3
5
1
8
4
6
2
Misting Nozzles
UV
Lamp
Water
Pump
Air Pump 2
Exhaust air vents
+ actuators
Exhaust air vents
+ actuators
DOWNDRAUGHT EVAPORATIVE COOLING
Water
Storage
146

PREVIOUS RESEARCH

PHDC

Passive and Hybrid Downdraught Cooling (PHDC), in different forms, is now technically viable in many parts of the world. This has been established through a combination of research projects and built examples around the world.

Research in this area which has been supported by the European Commission includes the research project on 'Passive Downdraught Evaporative Cooling in non-domestic buildings – PDEC' (JOR-3CT950078), and the project on 'Passive Downdraught Evaporative Cooling using porous ceramic evaporators – EVAPCOOL' (ENK6-CT-2000-00346).

A market assessment study of the potential application of PDEC to both existing and new non-domestic buildings was completed in 2003 as part of the ALTENER II project on 'Solar Passive Heating and Cooling' (4.1030/C/00-009/2000).

PDEC. THE APPLICATION OF PASSIVE DOWNDRAUGHT EVAPORATIVE COOLING TO NON-DOMESTIC BUILDINGS, 1993 – 1996

Research project funded by the european commission

Contract no.: JO_R3-CT95-0078

Partners

De Montfort University, Leicester, UK
Short Ford Associates, London, UK
Mario Cucinella Architects, Paris, France
ESII, University of Malaga, Malaga, Spain
Conphoebus, Catania, italy
Microlide, St Leonard de Noblat, France

Background

Passive downdraught evaporative cooling (PDEC) involves spraying controlled volumes of microscopic water droplets into hot, dry ambient air, thereby causing it to cool and descend into a capture zone within a building. From here the cool air enters the adjacent occupied spaces through carefully sized and controlled openings. Close collaboration is needed between architects and engineers for PDEC to work well.

The European Commission supported an international team of architects, engineers and industrialists from four European countries to carry out research into the application of PDEC in non-domestic buildings.

Objectives of the research

1. To produce detailed architectural designs for one real refurbishment scheme and two new-build projects to illustrate the use of PDEC on urban sites in Southern Europe.

2. To identify the most appropriate computer algorithms and modelling methods for simulating PDEC and, if necessary, to improve these.

3. To assess the three PDEC buildings using state-of-the-art thermal simulation and computational fluid dynamics programs to determine: the quality of the internal environment; the energy savings compared to conventional equivalents; and the wider environmental benefits.

4. To produce detailed information about the mechanical and structural engineering implications of using PDEC and to quantify the costs relative to conventional buildings.

5. To conduct full-scale physical experiments in Sicily during two summer periods, and thus to understand by monitoring: how PDEC buildings operate; the quality of the internal environment which is produced; and how best to install, control and maintain PDEC equipment.

6. To map out the zone of applicability of the PDEC technique (within S. Europe and adjacent regions) using thermal simulation.

7. To produce a PDEC design guide targeted at architects, energy consultants and engineers which explains: how to design PDEC buildings; how to undertake credible simulations of their performance; and their engineering and cost implications using primarily illustrations and examples generated within the project.

Research activities

The research work of the partners was divided into four main areas: Architectural Design Studies (ADS), Building Performance Assessment (BPA), Experimentation and Monitoring (E&M).

The Architectural Design studies included studing a

reference building undergoing refurbishment in Seville, the design and construction of a full scale experimental building in Catania with a microniser tower, and the design of two new PDEC office buildings in Seville and Catania.

The Building Performance team used airflow simulations to compare the thermal environment, energy performance and cost of the new building design against conventional equivalents. Quantitative assessments were undertaken and airflow analysis provided a clear understanding of internal comfort conditions. These analyses established the optimal PDEC operation and control strategies for the design.

Full scale field monitoring of the experimental building was undertaken at Conphoebus during two summer campaigns.

Results and Conclusions

Architectural Design.

In architectural and engineering terms, it has been demonstrated that a mixed-mode approach (PDEC plus supplementary cooling provision) is technically and economically viable, and is competitive against conventional air-conditioning for Southern Europe.

The exploration of PDEC design has developed into an architectural language that can bridge the gap between historical and contemporary design.

Building Performance

For the climates and building types considered, a mixed-mode solution is most appropriate. This involves natural ventilation at night and PDEC during the day with, if temperatures are still uncomfortably high, mechanical cooling. PDEC alone could not generate a sufficient temperature depression to achieve comfort in all ambient conditions. Whilst no new software has been written, a clear performance analysis methodology, using thermal and computational fluid dynamics simulation, has been developed for assessing PDEC in new and refurbished buildings. Maps of Europe indicating the applicability of PDEC have been generated.

Experimentation

It has been possible to collect and analyse performance data from a PDEC experimental test facility in Catania over two consecutive summer periods. Data sets for a range of microniser operating conditions have been collected and used to verify the simulation techniques employed.

PDEC can avoid the need for ductwork, fans and suspended ceilings, thereby reducing the overall height of buildings. In less severe climates, or in less demanding buildings, PDEC might avoid the need for air-conditioning entirely. PDEC provides 100% fresh, cool air with no recirculation – as is often the case in air-conditioned buildings. PDEC ought, therefore, to provide an environment with high air quality.

The results of the research are most encouraging and have demonstrated that PDEC is technically and economically viable. With careful attention to design detail, cooling energy savings of up to 66% for a modern office building were predicted. The technical and commercial potential of PDEC is very promising. Initial estimates suggest, with modest uptake, energy savings in Spain and Italy could be around 12.5 million kWh per year. The results of the regional performance assessment suggest a very large potential market for this novel technology.

EVAPCOOL. PASSIVE DOWNDRAUGHT COOLING SYSTEMS USING POROUS CERAMIC EVAPORATORS 1993 – 1996

Research project funded by the European Commission under the Fifth Framework Programme

Contract no.: ENK6-CT-2000-00346
Project No.: NNE5-2000-00511

Partners

WSP Environmental LTD, UK (Coordinator)
University of Nottingham, UK
Axima, Switzerland

Background

This project was preceded by the JOULE Research Project into Passive Downdraught Evaporative Cooling (PDEC 1996-1999) which concluded that in architectural and engineering terms a mixed mode approach (PDEC + back-up system) is technically and economically viable, and is competitive against conventional air-conditioning for office buildings in southern Europe. This research was based on a specification using misting nozzles under high pressures (20-50bar) to generate a high rate of evaporation and to minimise moisture carry-over. This system provides improved performance and other advantages over directly wetted cellulose pads, but it still has a number of disadvantages: high quality water required, risk of nozzles blocking up, risk of dripping (carry-over), risk of microbiological contamination and other maintenance issues.

It is clearly desirable to design a cooling system which avoids these problems. Alternative systems using porous media can overcome many of these disadvantages and would also extend the geographical and climatic range of applications.

Objectives of the research

1. To develop theoretical models of downdraught cooling resulting from evaporation through porous ceramic panels in both direct and indirect systems.
2. To design and develop new products for improved evaporator performance as part of a passive downdraught cooling system. Specifically, to design, manufacture and test two innovative systems incorporating porous ceramic evaporators.
3. To compare the performance of the direct and indirect evaporative cooling systems with conventional chilled water cooling coils under buoyancy driven airflow.
4. To investigate the architectural and engineering implications of integrating these new passive downdraught cooling systems and components in case study buildings.

Research activities

The research work concerned three themes: theory, design and manufacturing.

Theory

- *Theoretical Models – Axima*

Review of the theoretical basis for evaporative cooling through porous ceramic materials and the development of dynamic modelling tools to enable performance assessment of the technique applied to non-domestic buildings in both Northern and Southern Europe.

- *Laboratory Scale Measurements - Nottingham University.*

Design and build a laboratory scale prototype for two porous ceramic component/system ideas and to test each system with measurements under controlled conditions of temperature, relative humidity and buoyancy air flow.

Design

- Integrated System Design - WSP Environmental

Investigate the architectural and engineering design implications in terms of detailed design, performance and cost, of the proposed passive downdraught cooling systems and their integration into the context of a real building. Investigate the manufacturing options and limitations on the design of the prototype.

- System Performance Analysis – Axima

Evaluate the performance of a case study office building with a passive downdraught cooling system incorporating either direct or indirect ceramic evaporators and compare this performance and cost with that of the same building with chilled water cooling coils.

Manufacture

– Prototype Design & Manufacture – WSP Environmental

Design and manufacture a full scale prototype of the proposed passive downdraught cooling systems.

– Prototype Installation & Testing – Nottingham Uni.

Install and test the full scale prototype systems in an existing climate chamber in Nottingham.

Results and Conclusions

During the 30 months research project the following objectives were achieved:

1. Theoretical models of downdraught cooling resulting from evaporation through porous ceramic panels in both direct and indirect systems were developed.

2. The design and development of new products for improved performance of passive downdraught cooling systems were undertaken. Specifically, two innovative direct evaporative systems incorporating porous ceramic evaporators were designed, manufactured and tested.

3. A comparison between the performance of the direct and indirect evaporative cooling systems with conventional chilled water cooling coils under buoyancy driven airflow was made.

4. The architectural and engineering implications of integrating these new passive downdraught cooling systems and components in a case study building were investigated.

ALTENER. SOLAR PASSIVE HEATING AND COOLING MARKET ASSESSMENT OF THE POTENTIAL APPLICATION OF PASSIVE DOWNDRAUGHT EVAPORATIVE COOLING IN SOUTHERN EUROPE 2001 - 2003

Research project funded by the European Commission

Contract no.: 4.1030/C/00-009/2000
Part 1, Cluster 9

Partners

WSP Environmental Ltd., London, UK
Mario Cucinella Architects, Bologna, Italy
Davis Langdon Consultancy, London, UK
AICIA-University fo Seville, Seville, Spain

Background

Part 1 of Cluster 9 builds on work from the JOULE II R&D project, which has demonstrated the technical and economic viability of the Passive Downdraught Evaporative Cooling (PDEC) technique applied to new office and commercial buildings.

The purpose of Part 1 was to undertake a market assessment of the potential for the application of Passive Downdraught Evaporative Cooling (PDEC) to new and existing non-domestic buildings in Southern Europe. PDEC is a technique which potentially could become a substitute (in whole or part) for conventional air-conditioning.

Objectives of the research

1. To establish a database on existing and new non-domestic buildings in Italy, Spain, Portugal and Greece.
2. To estimate the proportion of existing buildings and new construction to which PDEC is applicable and the possible cost and energy savings in such buildings by implementing suitable measures over a period of time.
3. To determine those climatic regions in Southern Europe where PDEC is suitable and to overlay the location and size (population) of settlements on maps of PDEC applicability.
4. To predict the likely market penetration over the next 30 years within Italy, Spain, Portugal and Greece.
5. To indicate where promotional effort needs to be placed, to publish the study findings and disseminate results to key groups within the EU.

Research activities

This project followed an approach developed by D. M. Lush & J. L. Meikle which is based on an examination of building categories and stock characteristics, and the estimation of the building categories where potential for energy savings exist. Potential was gauged against a number of parameters which affect applicability (e.g. for PDEC, the existence of a transitional space, atrium or courtyard). These parameters were then ranked according to an assessment of their impact, with regard to energy savings and cost-effectiveness.

Seven discrete tasks were completed, each of which has contributed to the assessment of technical applicability, market barriers, cost benefits and market penetration.

1. Pilot Exercise / Stock and Investment Assessment
2. Typological Evaluation (Urban Morphology)
3. Mapping
4. Market Barriers to PDEC
5. Cost Benefit Evaluation
6. Market Penetration
7. Review & Dissemination

Results and Conclusions

Typological Evaluation

Studies of building typology and morphology have been undertaken as part of the process of assessing the applicability of PDEC to non-domestic buildings in southern Europe. A preliminary generalised benchmark, which relates PDEC applicability to the void/floor ratio, has been derived from the study of precedents.

Analysis of sample areas of predominantly commercial buildings in major cities in Spain, Italy and Portugal has revealed the proportion of the building stock which is suitable for different levels of PDEC intervention. The study indicates that, for all the cities studied, between 62% and 82% of all buildings are suitable for the application of PDEC.

Mapping

A new approach has been developed, based on the current developed mapping methodology, and using simulation tools, in order to achieve customisable charts, for specific buildings, systems and control strategies.

Market Barriers to PDEC

In terms of impact, barriers can be divided into those that cannot readily be overcome and those that can. The principal barriers that cannot readily be overcome are: technical barriers and regulatory and institutional barriers Other barriers are economic and financial Information, awareness and capacity

Cost Benefit Evaluation

A spreadsheet cost model was constructed to calculate the cost of providing an intervention using the data from the surveys and savings were calculated on the basis of the stock depth profile. The table below summarises the stock areas and potential energy savings and shows that energy savings could be around 1.5% to 2.5% of the national annual electricity consumption.

COUNTRY	Commercial buildings area millions m2	National electricity consumption millions KWh	PDEC potencial energy savings millions KWh	Energy saving as % of national electricity consumption	Reduction in CO2 emissions tonnes pa	Potential value of energy saving pa euro million
GREECE	39	46,099	**1,124**	2.44%	**766,596**	**85.4**
SPAIN	116	201,159	**3,341**	1.66%	**1,472,654**	**257.9**
ITALY	161	283,737	**4,637**	1.63%	**2,809,643**	**490.4**
PORTUGAL	25	41,146	**720**	1.75%	**391,414**	**59.0**

Market Penetration

This task produced estimates of the likely penetration of PDEC applications in both new and refurbished commercial buildings in Greece, Italy, Portugal and Spain in the period 2005 to 2025. Typically, buildings represent around 50% of national energy consumption and commercial buildings may represent 10 – 20% of that. It is likely, therefore, that the contribution of PDEC to savings in the energy consumption by, and the CO2 emissions from, commercial buildings in the four countries are likely to be of the order of 0.5 to 1.0% in 2005 rising to 8 to 15% in 2025.

REFERENCES

Chapter 1

1 Meteotest, ***Meteonorm - Global Solar Radiation Database METEONORM***, 2008, http://www.meteonorm.com/pages/en/meteonorm.php.

2 Architecture and Urbanism Design UCLA, ***Energy Design Tools***, 2009, http://www.energy-design-tools.aud.ucla.edu/.

3 M. Santamouris, ***Energy and Climate in the Urban Built Environment***, London: James & James, 2001. ISBN: 1873936907

4 Edward Ng, ***Designing for Thermal Comfort and Air Ventilation in High Density City Conditions – Air Ventilation Assessment System (AVAS)***, The Hong Kong Institute of Architects Journal (2007): 24-31.

5 Arthur Rosenfeld et al., ***Painting the Town White-- and Green***, MIT's Technology Review, 1997, http://www.sfrc.ufl.edu/urbanforestry/Resources/PDF%20downloads/Rosenfeld_1997.pdf.

6 Khandaker Shabbir Ahmed, ***Comfort in urban spaces: defining the boundaries of outdoor thermal comfort for the tropical urban environments***, Energy and Buildings 35, no. 1 (January 2003): 103-110. ISBN: 0378-7788

7 Ibid.

8 Hassan Fathy, ***Natural Energy and Vernacular Architecture: Principles and Examples with Reference to Hot Arid Climates***, University Of Chicago Press, 1986. ISBN: 0226239179

9 Battle McCarthy, ***Wind Towers (Detail in Building)***, John Wiley & Sons, 1999. ISBN: 0471980870

10 Malcolm Orme, M Liddament, and A Wilson, ***Coventry: Air Infiltration and Ventilation Centre***, 1998, http://openlibrary.org/b/OL18465873M/Numerical_data_for_air_infiltration_and_natural_ventilation_calculations.

11 B. Givoni, ***Building design principles for hot humid regions***, Renewable Energy 5, no. 5-8 (August 1994): 908-916.

12 Brian Ford and K Cairns, ***Market Assessment of passive downdraught evaporative cooling in non-domestic buildings in Southern Europe***, in, vol. 2. Proceedings from the 3rd European Conference on Energy Performance & Indoor Climate in Buildings, EPIC, Lyon, France, 2002, 505-510.

13 CIBSE, ***Natural Ventilation in Non-Domestic Buildings***, Applications Manual No. 10, London: Chartered Institution of Building Services Engineers, 2005. ISBN: 1 903287 561

[14] Simos Yannas, Evyatar Erell, and Jose Luis Molina, ***Roof Cooling Techniques a Design Handbook***, London: Earthscan, 2006.

[15] T. R Oke, ***Boundary Layer Climates***, 2nd ed., London: Methuen, 1987. ISBN: 1 903287 561

[16] H Mulligan, ***Environmental Characteristics of the Vernacular Underground Dwelling***, in Dwellings. Proceedings from the PLEA Conference, Crete, Greece, 1983.

[17] Baruch Givoni, ***Passive and Low Energy Cooling of Buildings***, New York: John Wiley, 1994. ISBN: 0471284734

[18] N.T. Bowman et al., ***Passive Downdraught Evaporative Cooling: I. Concept and Precedents***, Indoor and Built Environment 9, no. 5 (September 1, 2000): 284-290.

[19] W.A. Cunningham and T.L. Thompson, ***Passive Cooling with Natural Draft Cooling Towers in combination with Solar Chimney***. Proceedings from the PLEA Conference, Hungary. The Hungarian Society of Sciences, 1989.

[20] B. Givoni, ***Performance and applicability of passive and low-energy cooling systems***, Energy and Buildings 17, no. 3 (1991): 177-199.

[21] Hassan Fathy, ***Natural Energy and Vernacular Architecture: Principles and Examples with Reference to Hot Arid Climates***, University Of Chicago Press, 1986. ISBN: 0226239179

[22] N Al Hemiddi and B Givoni, ***Applicability of a 'Shower' Passive Cooling Tower in Hot Dry Climate***, in Proceedings from the 20th National Passive Solar Annual Conference, Minneapolis, Minnesota - U.S.A., 1995, 143-148.

[23] N Al Hemiddi, ***Preliminary investigation of the effect of a passive direct evaporative cooling system on a courtyard at UCLA***, in Architecture and Urban Space. Proceedings from the 9th PLEA Conference, Seville, Spain: Pergamon, 1991, 643-648. ISBN: 9780792314189

[24] Y. Etzion et al., ***Adaptive architecture: integrating low-energy technologies for climate control in the desert***, Automation in Construction 6, no. 5-6 (September 1997): 417-425. ISSN: 0926-5805

[25] D. Pearlmutter et al., ***Refining the use of evaporation in an experimental down-draft cool tower***, Energy and Buildings 23, no. 3 (March 1996): 191-197. ISSN: 0378-7788

[26] S. Alvarez, E.A. Rodriguez, and J.L. Molina, ***The Avenue of Europe at Expo 92: Application of Cool Towers***, in Architecture and Urban Space. Proceedings from the 9th Passive and Low Energy Architecture, Seville, Spain: Pergamon, 1991, 195-202. ISBN: 9780792314189

[27] Brian Ford and G Papagiannopoulos, ***Evaporative Cooling Using Porous Ceramic Bricks: Experimental Results from Greece***, in Rethinking Development. Proceedings from the 20th Passive and Low Energy Architecture, Santiago, Chile, 2003.

[28] Rosa Schiano-Phan, ***The Development of Passive Downdraught Evaporative Cooling Systems Using Porous Ceramic Evaporators and their application in residential buildings***, Plea 2004 The 21st Conference on Passive & Low Energy Architecture, Eindhoven, Netherlands, 19-22 September 2004, (ISBN: 9038616368) Vol. 2, pp. 1249-1254.

[29] William JR Curtis, ***Reflections on Water: William JR Curtis reviews the highs and lows of Spain's aquatic Expo 2008, Architecture Today, September 1, 2008***, http://www.architecturetoday.co.uk/Articles/view.php?id=23893.

Chapter 2

[1] Brian Ford and K Cairns, ***Market Assessment of passive downdraught evaporative cooling in non-domestic buildings in Southern Europe***, in, vol. 2. Proceedings from the 3rd European Conference on Energy Performance & Indoor Climate in Buildings, EPIC, Lyon, France, 2002, 505-510.

[2] Brian Ford and Raul Moura, ***Market Assessment of passive downdraught evaporative cooling in non-domestic buildings in southern Europe***, Final Report, ALTENER II Project on 'Solar Passive Heating and Cooling', European Commission - DG Reseach, 2003, http://www.phdc.eu/uploads/media/ALTENER_1_Final_report_extract.pdf.

[3] N Bowman et al., ***Application of passive downdraught evaporative cooling (PDEC) to non-domestic buildings***, Renewable Energy 10, no. 2-3 (1997): 191-196. ISSN: 0960-1481

[4] Brian Ford, Camilo Diaz, and Mark Hewitt, ***Passive Downdraught Cooling: Architectural Integration in Seville***, in Architecture, City, Environment. Proceedings from the Passive and Low Energy Architecture, Cambridge, United Kingdom: James & James, 2000, 83-88. ISBN: 1902916166

[5] M.J. Cook et al., ***Passive Downdraught Evaporative Cooling: II. Airflow Modelling***, Indoor and Built Environment 9, no. 6 (November 1, 2000): 325-334.

[6] N.T. Bowman et al., ***Passive Downdraught Evaporative Cooling: I. Concept and Precedents***, Indoor and Built Environment 9, no. 5 (September 1, 2000): 284-290.

[7] M.J. Cook et al., ***Passive Downdraught Evaporative Cooling: II. Airflow Modelling***, Indoor and Built Environment 9, no. 6 (November 1, 2000): 325-334.

Chapter 3

[1] National Weather Service, ***Phoenix, AZ***, 2008, http://www.wrh.noaa.gov/psr/.

[2] Kevin Winschel, Interview, 2007.

[3] Environmental Protection Agency, ***National Action Plan for Energy Efficiency***, 2008, http://www.epa.gov/RDEE/energy-programs/napee/meetings/sector.html.

[4] Mahadev Raman, ***Notes from teleconference between Dr. Rosa Schiano-Phan and Mr. Mahadev Raman at Arup New York***, 2009.

[5] Kevin Winschel, Interview, 2007.

[6] Environmental Protection Agency, ***National Action Plan for Energy Efficiency***, 2008, http://www.epa.gov/RDEE/energy-programs/napee/meetings/sector.html.

[7] Mahadev Raman, ***Notes from teleconference between Dr. Rosa Schiano-Phan and Mr. Mahadev Raman at Arup New York***, 2009.

[8] Brian Ford et al., ***Cooling without air conditioning: The Torrent Research Centre, Ahmedabad, India***, Renewable Energy 15, no. 1-4 (1998): 177-182. ISSN: 0960-1481

[9] L.E. Thomas and G Baird, ***Post-occupancy evaluation of passive downdraft evaporative cooling and air-conditioned buildings at Torrent Research Centre, Ahmedabad, India***, in Proceedings of the 40th Annual Conference of the Architectural Science Association ANZAScA. 2006, 97-104. ISBN: 0-9586961-2-8

[10] Ibid.

[11] George Baird, ***The Architectural Expression of Environmental Control Systems***, London: Spon Press, 2001. ISBN: 0419244301

[12] Mili Majumdar, ***Energy Efficient Buildings in India***, Tata Energy Research Institute, TERI, 2001.

[13] Brian Ford et al., ***Cooling without air conditioning: The Torrent Research Centre, Ahmedabad, India***, Renewable Energy 15, no. 1-4 (1998): 177-182. ISSN: 0960-1481

[14] L.E. Thomas and G Baird, ***Post-occupancy evaluation of passive downdraft evaporative cooling and air-conditioned buildings at Torrent Research Centre, Ahmedabad, India***, in Proceedings of the 40th Annual Conference of the Architectural Science Association ANZAScA. 2006, 97-104. ISBN: 0-9586961-2-8.

[15] Inderjeet Singh and Axel Michaelowa, ***Indian Urban Building Sector: CDM Potential through Energy Efficiency in Electricity Consumption***. Hamburg Institute of International Economics, 2004, http://ideas.repec.org/p/ags/hiiedp/26323.html.

[16] L.E. Thomas and G Baird, ***Post-occupancy evaluation of passive downdraft evaporative cooling and air-conditioned buildings at Torrent Research Centre, Ahmedabad, India***, in Proceedings of the 40th Annual Conference of the Architectural Science Association ANZAScA. 2006, 97-104. ISBN: 0-9586961-2-8

[17] Ibid.

[18] Ibid.

[19] Mili Majumdar, ***Energy Efficient Buildings in India***, Tata Energy Research Institute, TERI, 2001. ISBN: 8185419825

[20] J. Fergus Nicol et al., ***Climatic variations in comfortable temperatures: the Pakistan projects***, Energy and Buildings 30, no. 3 (August 1999): 261-279.

[21] L.E. Thomas and G Baird, ***Post-occupancy evaluation of passive downdraft evaporative cooling and air-conditioned buildings at Torrent Research Centre, Ahmedabad, India***, in Proceedings of the 40th Annual Conference of the Architectural Science Association ANZAScA. 2006, 97-104. ISBN: 0-9586961-2-8

[22] Inderjeet Singh and Axel Michaelowa, Indian Urban Building Sector: CDM Potential through Energy Efficiency in Electricity Consumption. Hamburg Institute of International Economics, 2004, http://ideas.repec.org/p/ags/hiiedp/26323.html.

[23] Brian Ford and Camilo Diaz, ***Passive downdraft cooling: hybrid cooling in the Malta Stock Exchange, in Rethinking Development: Are We Producing a People Oriented Habitat?*** Proceedings from the PLEA Conference, Santiago, Chile, 2003.

Chapter 4

[1] ALTENER II, Solar Passive heating and Cooling. ***Market Assessment of the Potential Application of Passive Downdraught Evaporative Cooling in Southern Europe 2001 - 2003 - 4.1030/C/00-009/2000***, Part of final report, ALTENER II Project on Solar Passive Heating and Cooling, European Commission - DG Research, 2003.

[2] EVAPCOOL - ***Passive Down-draught Evaporative Cooling using porous ceramic evaporators*** (1998-2002). Contract No ENK6-CT-2000-00346. Project No NNE5-2000-00511, Final report, 2003.

[3] European Commission, *DG research Altener cluster 9 (4.1030/C/00-009/2000)* ***Solar Passive Heating and Cooling***, *Prof. Brian Ford and Raul Moura.*

[4] https://www.cia.gov/library/publications/the-world-factbook/index.html

[5] The SOLVENT *study found that these three were the principal barriers to the adoption of passive design in Greece.*

Chapter 5

[1] Baruch Givoni, ***Passive and Low Energy Cooling of Buildings***, New York: John Wiley, 1994. ISBN: 0471284734

[2] P. O. Fanger, ***Thermal Comfort: Analysis and Applications in Environmental Engineering***, Copenhagen: Danish Technical Press, 1970. ISBN: 8757103410

[3] M.A. Humphreys, ***Field studies of thermal comfort compared and applied,*** Building Services Engineering 44, no. 1 (1976): 5-23, 27; M.A. Humphreys, ***Outdoor temperatures and comfort indoors***, Batim Int Build Res Pract 6, no. 2 (1978): 92-105; J.F. Nicol and M.A. Humphreys, ***Thermal comfort as part of a self-regulating system***, Build Res Pract 1, no. 3 (1973): 174-179.

[4] Geun Young Yun, Paul Tuohy, and Koen Steemers, ***Thermal performance of a naturally ventilated building using a combined algorithm of probabilistic occupant behaviour and deterministic heat and mass balance models***, Energy and Buildings 41, no. 5 (May 2009): 489-499. ISSN: 0378-7788

[5] M.A. Humphreys, ***Thermal comfort, climate and buildings***, in, vol. 44. Proceedings from the 1992 World Renewable Energy Congress, London, UK, 1992, 5-23, 27
R.J. de Dear and G. Brager, ***Developing an adaptive model of thermal comfort and preference***, in ASHRAE Transactions, vol. 104 (ASHRAE, 1998), 27-49.

[6] ASHRAE, ***Thermal Environmental Conditions for Human Occupancy***, ANSI/ASHRAE standard*v55-2004 (Atlanta Georgia: American Society of Heating, Refrigerating and Air-Conditioning Engineers, 2004);
CIBSE, Comfort Standard; British Standards, ***Indoor environmental input parameters for design and assessment of energy performance of buildings addressing indoor air quality, thermal environment, lighting and acoustics***, BS EN 15251:2007 (BSI, 2008).

[7] K. J. Lomas et al., ***Building bioclimatic charts for non-domestic buildings and passive downdraught evaporative cooling***, Building and Environment 39, no. 6 (June 2004): 661-676. ISSN: 0360-1323

[8] W.A. Cunningham and T.L. Thompson, ***Passive Cooling with Natural Draft Cooling Towers in combination with Solar Chimney***, in Passive and Low Energy Housing. Proceedings from the PLEA Conference, Hungary: The Hungarian Society of Sciences, 1989.

[9] Brian Ford et al., ***Cooling without air conditioning: The Torrent Research Centre, Ahmedabad, India***, Renewable Energy 15, no. 1-4 (1998): 177-182. ISSN: 0960-1481.

[10] T. Thompson, Nader V. Chalfoun, and Martin R. Yoklic, ***Estimating the performance of natural draft evaporative coolers***, Energy Conversion and Management 35, no. 11 (November 1994): 909-915. ISSN: 0196-8904

[11] Baruch Givoni, ***Passive and Low Energy Cooling of Buildings***, New York: John Wiley, 1994. ISBN: 0471284734

[12] Benjamin Stein et al., ***Mechanical and Electrical Equipment for Buildings***, 11th ed. Hoboken, N.J: Wiley, 2009. ISBN: 9780470195659

[13] Alison G Kwok and Walter T Grondzik, ***The Green Studio Handbook: Environmental Strategies for Schematic Design***, Oxford: Architectural Press, 2007. ISBN: 0750680229

[14] Sprecher and Tillenkamp, ***Energy saving systems in building technology based on concrete-core-cooling***, International Journal of Ambient Energy 24, no. 1 (2003): 29-34.

[15] Rosa Schiano-Phan, ***The Development of Passive Downdraught Evaporative Cooling Systems Using Porous Ceramic Evaporators and their application in residential buildings***, in Built environments and environmental buildings, vol. 2. Proceedings from the 21st PLEA Conference, Netherlands, 2004, 1249-1254.

[16] Rosa Schiano-Phan, ***Environmental retrofit: building integrated cooling in housing***, in Architectural Research Quarterly, Volume 14, Issue 1, March 2010, Cambridge University Press. ISBN 13591355.

Chapter 6

[1] J.M. Salmeron et al., ***Flow Pattern Effects on Night Cooling Ventilation*** 6, no. 1 (2007).

[2] S. Alvarez, J.R. Maestre, and R. Velazquez, ***Design Methodology and Cooling Potential of the Environmental Heat Sinks***, International Journal of Solar Energy 19 (1997): 179-197.

Chapter 7

[1] W Bordass, A Leaman, and S Willis, ***Control Strategies for Building Services: the role of the user***, in Buildings and the Environment. Proceedings from the CIB/BRE Conference, Garston, Building Research Establishment, 1994.

[2] CIBSE, ***Natural Ventilation in Non-Domestic Buildings***, Applications Manual No. 10, London: Chartered Institution of Building Services Engineers, 2005. ISBN: 1 903287 561

[3] Ibid.

[4] Ibid.

[5] D. Pearlmutter et al., ***Refining the use of evaporation in an experimental down-draft cool tower***, Energy and Buildings 23, no. 3 (March 1996): 191-197. ISSN: 0378-7788

[6] Ibid.

[7] Brian Ford and Camilo Diaz, ***Passive downdraught cooling: hybrid cooling in the Malta Stock Exchange***, in Rethinking Development: Are We Producing a People Oriented Habitat? Proceedings from the PLEA Conference, Santiago, Chile, 2003.

[8] Standards Australia, AS/NZS 3666 Set:2006. ***Air-handling and water systems of buildings***. Australia, New Zealand, 2006, http://www.standards.com.au.

FIGURES

Chapter 2

Chapter 3

Chapter 4

Chapter 5

vegetarian barbecue cookbook

vegetarian barbecue cookbook

Mary Gwynn

MEREHURST

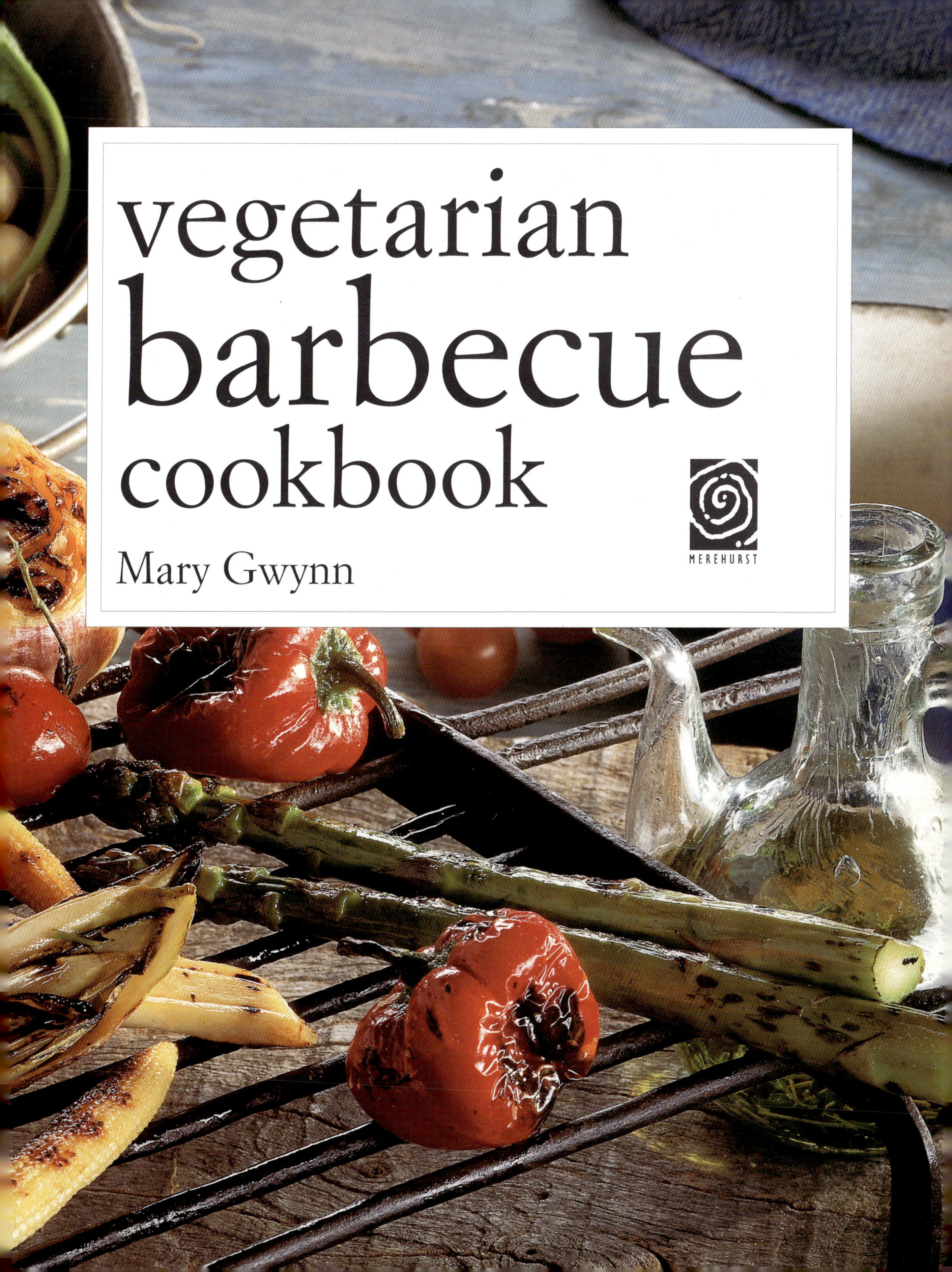

First published 1998 by Merehurst Limited
Ferry House, 51-57 Lacy Road, Putney, London SW15 1PR

ISBN 1-85391-678-1

A catalogue record for this book is available from the British Library.

Editor: Bridget Jones
Design: Hammond Hammond
Photographer: Ken Field
Home Economist: Louise Pickford
Stylist: Suzy Gittins

Colour separation by Bright Arts
Printed in Italy by Olivotto

NOTES ON USING THE RECIPES

- Spoons – a standard spoon measurement is used in all recipes
  teaspoon = one 5 ml spoon
  tablespoon = one 15 ml spoon
  All spoon measures are rounded.
- All eggs are size 3 unless otherwise specified.
- Ovens should be preheated to the specified temperature.
- For all recipes, quantities are given in metric and Imperial. Follow one set of measures but not a mixture as they are not interchangeable.
- The recipes in this book have been devised so that they can be prepared in the time it takes for the coals to heat up to the correct cooking temperature from a cold start. Some foods need marinating, so check ahead when planning a barbecue.

CONTENTS

FOREWORD

I first discovered the pleasures of impromptu meals as a child and I have enjoyed eating outdoors ever since. Those totally unsophisticated childhood barbecues were normally instigated by my father, during outings in the countryside on cold wintry days. The menu invariably consisted of small baked potatoes, wrapped in foil and cooked among the hot ashes of a fire made of wood we had collected, and when they were cool enough to handle, the potatoes were still wonderfully warming in small, chilly hands. Of course, there were baked beans – they were heated in the opened can and served on toasted currant buns. This may sound an odd combination and it had its detractors at the time, but to me it was totally satisfying in terms of both food and setting. In fact, those adventures sum up the joys of cooking and eating outdoors, when the food seems to taste better as everyone relaxes and joins in the informal fun.

Since then, I have experienced barbecues around the world. They range from riotous gatherings on a beach in the South Pacific, where we cooked a real Australian 'barbie' on vast grills set over oil drums, to the 'real thing' on my first trip to the United States of America, when friends in Connecticut cooked out on the barbecue by the pool and we ate char-grilled corn-on-the-cob with the biggest steaks that I had ever seen.

In those days, barbecued food offered little other than steaks, hamburgers and sausages, which, unless treated with some care, ended up charred to a crisp on the outside and unpleasantly raw in the centre. Now, however, with the huge interest in Mediterranean and Asian food, focused on sealing in flavours by grilling, roasting and searing, and the move away from meat-based meals, the barbecue has been liberated as a healthy, innovative method of cooking that also offers all kinds of possibilities for vegetarians.

The recipes in this book offer only an idea of the versatility of the barbecue for meat-free eating and they are intended as inspiration for everyone who enjoys good food. I hope you enjoy cooking and eating them as much as I have.

Mary Gwynn

GETTING STARTED

WHICH BARBECUE?

All grills are based on the same simple structure, with a bed for fuel and a metal cooking rack set over it. Beyond this, there is an ever-increasing range of barbecues and selecting the most suitable can be confusing.

First, consider how often you will use your barbecue; how many people you are likely to feed; and where you intend to cook. This should enable you to choose between the state-of-the-art kettle barbecues or hooded grills at one end of the scale and the portable or tabletop hibachi-type grills.

Whichever barbecue you choose, make sure it is sturdy, with a cooking rack that is easy to remove or re-position.

Braziers or Open Grills

These are portable and free-standing, often with wheels or detachable legs. It is essential to look for a grill that is sturdy, with strong, solid legs which will not topple easily.

Some can be quite low, which can be painful on the backs of any tall chefs. Many come with windshields to shelter the coals and conserve some of the heat. Pick one with adjustable settings for the cooking rack and air vents that can be opened or closed, providing some degree of heat control.

This type is best for quick-cooking foods, no more than 4 cm (1½ inch) thick.

Kettle Grills

These covered grills have adjustable hoods which allow for three cooking options: as an open grill, covered oven or smoker. When closed, the heat is reflected to speed up the cooking process and make it more even, which is ideal for cooking thicker foods or whole vegetables. Add aromatic herbs or wood chips for a smoked flavour.

The cooking grid is not usually adjustable, so it is difficult to control the temperature, and the lid is not hinged, so it can be hot and unwieldy to handle.

Portable Grills

The most common is the hibachi grill. Named after a Japanese word meaning 'fire box', these are small, inexpensive, easy to carry and ideal for camping or picnics.

Other portable barbecues are readily available, with collapsible legs and carrying handles, and extras, such as windshields or adjustable racks. Disposable foil barbecues that burn for an hour or so are also sold in most supermarkets, but they are only suitable for cooking a few items.

Gas and Electric Grills

Quick to heat and easy to use, these are designed along the same lines as kettle or hooded barbecues. However, you will miss the authentic char-grilled flavour.

Built-in Barbecues

You can build your own barbecue on a patio or sheltered area away from the house. Do-it-yourself books offer step-by-step instructions; complete packs of materials and instructions are available.

SELECTING THE FUEL

I prefer hardwood charcoal as it burns hotter and cleaner than wood, and provides uniform heat. Good-quality briquettes will do, but you must wait until the coals have burnt down until they have a coating of grey ash as they contain chemical binders and fillers that can, if not completely burnt off, give the food an unpleasant taste.

Packs of aromatic wood chips, such as hickory and maple, can be added for a smoky flavour, or you can add woody herbs such as fennel, rosemary and sage.

LIGHTING THE BARBECUE

You will need a 2.5 – 5 cm (1 – 2 inch) base of charcoal or wood under the cooking area; remember, if you want to cook for longer than an hour, you will have to add more charcoal during cooking.

A useful tip is to line the barbecue base with heavy-duty foil to reflect the heat back up at the food and this also makes the barbecue easier to clean afterwards.

You can use paper and kindling, chemical blocks or fluid fire lighters to light the barbecue. If using chemical lighters or fluid, make sure they have burnt off before you start cooking or the food will have a chemical taste. For paper and kindling, make a bed of crumpled newspaper and lay bits of kindling on top. Pile the coals into a pyramid and light the paper or fire lighter.

Cooking Temperature

Leave the barbecue to burn for 30 to 40 minutes, until the flames subside and the coals are covered in a fine white ash, before starting to cook. The coals should glow red in patches to give the medium-hot heat which is perfect for cooking vegetables.

To top up the barbecue if the fire has burnt down too far, arrange more fuel around the edge and leave it to heat up before raking it into the centre, a few pieces at a time. Allow 15 minutes for the additional fuel to heat, with hood and vents open. Do not add cold fuel to the fire as it reduces the heat and slows the cooking.

COOKING TIMES

A charcoal barbecue and electric grill were used for testing. Times vary widely according to size and type of grill, weather and quantities cooked. Never leave the barbecue unattended and check food often.

COOKING TIPS

■ Most vegetables cook more evenly on a covered grill over medium heat. Pierce them with a wooden skewer to ensure they are tender all the way through.
■ Brush the grid with oil before and during use to prevent food from sticking.
■ I use long-handled hinged grill baskets to cook vegetables. They hold them in place and are easy to turn. Oil them well.
■ If cooking small or delicate vegetables on the grill, line a small area with heavy-duty foil and pierce it. This way, the vegetables will not fall onto the coals.
■ When cooking food wrapped in foil parcels, use a skewer to pierce the foil in several places to let the flavour through.
■ Wooden skewers are ideal for vegetables as they do not split delicate items, such as mushrooms, as much as metal ones can.
■ Soak wooden skewers in cold water for 15 minutes to prevent them from burning.
■ Use metal skewers for root vegetables, as they conduct heat to the centre well and promote quick, even cooking.
■ For even cooking, make sure all food is at room temperature before barbecuing.
■ Keep a water spray to hand, to damp down any flare ups.
■ Always keep the grill clean by scrubbing it with a wire brush after each use. Keep it freshly oiled, but free of stale grease.
■ At the end, turn off a gas or electric barbecue, cover and close down vents. If you are using an open grill, douse the coals lightly with water.

MENUS

With the exception of simple salads and plain rice, all the recipes for these menus are featured in the chapters which follow. The menus show just how imaginative barbecue meals can be, to suit all tastes and occasions. Page references for all the recipes are given in the index.

■ A SIMPLE SPRING BARBECUE
Char-grilled Asparagus with
Parmesan and Pine Nut Dressing
Mixed Vegetable Platter with
Chilli Mayonnaise
Garlic Rosemary Pizza Bread
A simple mixed leaf salad
Glazed Pears on Cinnamon Raisin Toast

■ A SUMMER FAMILY GATHERING
Roasted Tomato Crostini
Mushroom Carrot Burgers
Vegetable Satay
Grilled Fennel and Courgette Rice
Rocket, Red Pepper and
Roasted Shallot Salad
Banana and Strawberry Kebabs
with Toffee Sauce

■ A MIDSUMMER PATIO DINNER
Baby Artichokes and Aubergines with Alioli
Aubergine and Mozzarella Rolls
New Potato and Shallot Kebabs with
Coriander Pesto
Char-grilled Patty Pans with
Lemon Mustard Dressing
Toasted Bread Salad with Parmesan
Barbecued Peaches with
Whisky Mac Cream

■ AN AUTUMN BARBECUE
Garlic Mushrooms
Pumpkin and Celeriac Kebabs with
Cinnamon Butter
Vine-leaf Parcels with Feta and Olives
Roasted Plum Tomatoes with Chick Peas
Potato and Onion Bread
Figs with Honey and Lemon Grass

■ A MIDDLE-EASTERN INSPIRED BEACH BARBECUE
Herbed Cheese with Char-grilled Vegetables
Felafel in Pitta Pockets
Sweet Potatoes with Quick Curry Paste
Couscous with Barbecued Vegetables
Cucumber Raita
Naan or bought Indian flat bread
Exotic Fruit Kebabs with Cardamom Butter

■ A CASUAL BARBECUE
Roasted Gazpacho
Tomato and Avocado Salsa
Tortillas Stuffed with
Sweetcorn and Avocado
Barbecued Halloumi Strips
Patatas Bravas
A tomato and red onion salad
Peach and Blueberry Parcels

■ AN ASIAN-INSPIRED BARBECUE
Roasted Aubergine and Coriander Pâté
Sweetcorn and Red Onion Kebabs
with Sesame Oil
Thai Spiced Mushroom Kebabs
Peking-style Tofu with Pancakes
Cooked Thai jasmine rice
Ginger Maple Syrup Ice Cream with
Exotic fresh fruit salad

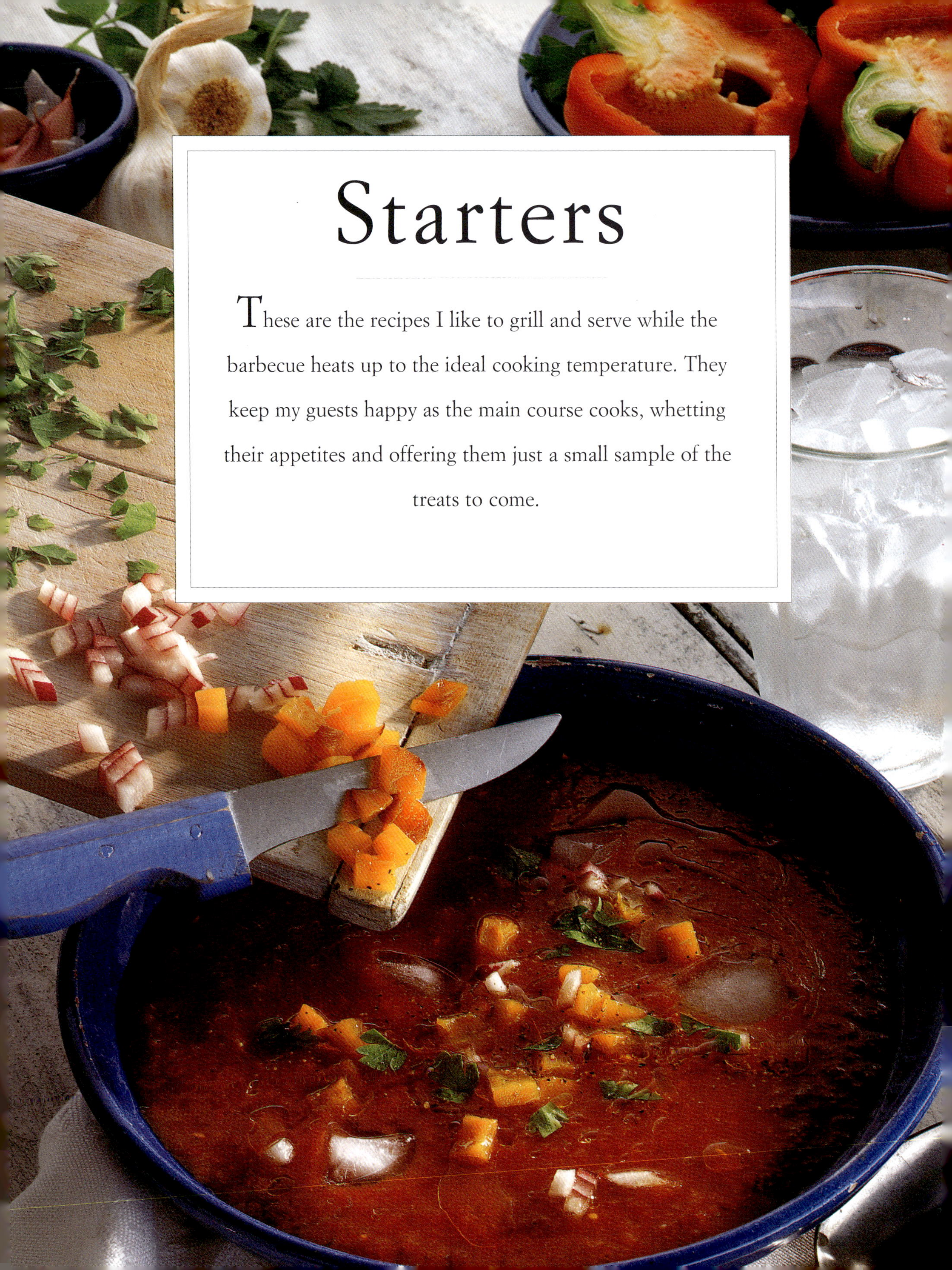

Starters

These are the recipes I like to grill and serve while the barbecue heats up to the ideal cooking temperature. They keep my guests happy as the main course cooks, whetting their appetites and offering them just a small sample of the treats to come.

CHAR-GRILLED TOMATO AND PEPPER SOUP

Time to prepare: 15 minutes
Time to cook: about 10 minutes

Serves 4

2 red peppers
1 green pepper
1 yellow pepper
500 g (1 lb) plum tomatoes
1 large mild onion, halved horizontally
2 tablespoons red wine vinegar
1 garlic clove
315 ml (10 fl oz) iced water
salt and freshly ground black pepper
ice cubes, to serve

Garnish
diced red onion
chopped fresh flat-leaf parsley

Gazpacho is a firm favourite with my daughters, who request it whatever the season and temperature outside. This is a version that I developed during a visit to Spain, where our holiday villa had a wonderful pool-side barbecue that we put to full use every evening. Other than a brush of the local olive oil to keep them moist as they grilled, intensely flavoured vegetables from the local market needed little enhancement. My children soon discovered the pleasures of eating aubergines and courgettes and they have been eating them ever since, so I can recommend this method of cooking to tempt unadventurous children into trying hitherto unpopular vegetables.

1 Place all the vegetables on the barbecue and cook for about 10 minutes, turning regularly, until the skins are blackened. Place the peppers and tomatoes in a plastic bag and leave for 5 minutes for the skins to loosen.
2 Peel the onion and coarsely chop the flesh. Skin and seed the peppers, then set the yellow pepper aside. Skin the tomatoes. Place the flesh of the red and green peppers in a blender with the onion and tomatoes. Process until smooth, then add the vinegar, garlic and water, and process again. Season to taste.
3 To serve, place two or three ice cubes in each bowl and pour in the soup. Finely chop the flesh of the reserved yellow pepper and scatter a little over each bowlful of soup. Sprinkle with diced red onion and chopped parsley to garnish and serve.

Cook's Tip
Sieve the soup to remove the tomato seeds for a smoother result. If you make the soup in advance, chill it and then stir it just before serving.

Adding the finishing touches to Char-grilled Tomato and Pepper Soup

ROASTED TOMATO CROSTINI

Time to prepare: 10 minutes
Time to cook: 10 minutes

Serves 4

250 g (8 oz) cherry tomatoes
2 garlic cloves, skins on
1 small red onion, halved
4 large fresh basil leaves, shredded
$^{1}/_{2}$ medium French baguette
4 tablespoons extra virgin olive oil
salt and freshly ground black pepper

Barbecuing or roasting is a wonderful way of giving flavour to the rather tasteless specimens of tomato that come out of many of our supermarkets – even those that come with the hopeful label 'grown for flavour'. Outside the short summer season when the tomato plants in my garden perform and produce fruit, I tend to buy cherry tomatoes when I want a really full tomato flavour for my dishes; they work well in this classic topping for crostini – Italian for 'little toasts'.

1 Place the whole tomatoes and garlic on the barbecue with the onion halves. Cook for about 5 minutes, turning occasionally, until just starting to blacken.
2 Skin the tomatoes and onion, chop the flesh, then place together in a small bowl. Add the basil and season well. Skin and mash the garlic in a separate small bowl.
3 Thickly slice the bread and toast it on the barbecue until both sides are golden. Drizzle olive oil on one side of each slice and spread with garlic. Top with the tomato mixture and serve.

CHARRED SPRING ONION BRUSCHETTA

Time to prepare: 15 minutes
Time to cook: 10 to 15 minutes

Serves 4

20 large spring onions
1 red chilli
1 tablespoon sesame oil
1 tablespoon soy sauce
1 tablespoon clear honey
2 tablespoons chopped fresh coriander
4 thick slices country-style bread
1 garlic clove

I really love bruschetta, the Italian combination of an open sandwich and garlic bread that makes a superb appetizer or snack lunch. In its elemental form, it is simply toasted bread rubbed with a cut clove of garlic and drizzled with olive oil (the best available). It can also be topped with all kinds of wonderful ingredients, and barbecued vegetables are among the most successful. Spring onions brushed with a little soy sauce and sesame oil, then cooked to a mellow, caramel-complemented tenderness, bring an oriental twist to this Mediterranean favourite. Try to find extra large spring onions as they are easier to handle and do not fall apart on the barbecue.

1 Cut the spring onions in half horizontally. Seed and finely chop the chilli. Mix the sesame oil, soy sauce and honey and stir in the chilli.
2 Place the spring onions on the barbecue and brush with the soy mixture. Cook for 10 to 15 minutes, turning and brushing with the soy sauce mixture, until starting to blacken. Remove the spring onions from the barbecue and cut them into 5 cm (2 inch) lengths. Mix the spring onions with the remaining marinade and coriander.
3 Toast the bread on the barbecue. Cut the garlic in half. Rub the toast with the cut surface of the garlic and pile the spring onions on top. Serve immediately.

Cook's Tip
For a more substantial salad gently stir cubes of barbecued tofu into the warm spring onions and leave to cool.

BABY ARTICHOKES AND AUBERGINES WITH ALIOLI

Time to prepare: 15 minutes
Time to cook: 10 minutes

Serves 4

250 g (8 oz) young baby artichokes (the type that are eaten whole)
250 g (8 oz) baby aubergines
extra virgin olive oil, for brushing

Alioli
5 garlic cloves
salt and freshly ground black pepper
1 egg
1 tablespoon lemon juice
155 ml (5 fl oz) Spanish extra virgin olive oil
lemon wedges, to garnish

One of my chief memories of visiting my sister Lucy when she spent a summer working on the Balearic island of Ibiza, is sitting at a beach-side restaurant, drinking sangria by the jugful and dipping bread into an incredibly powerful garlic sauce, known locally as alioli. The Spanish version of the aioli found in Southern France, it was definitely only for garlic lovers. On consulting the chef we discovered that ten cloves of garlic were added for every cup of oil! My version cuts the garlic by half, which seems a happy balance for barbecued vegetables. Just occasionally, I serve the full-strength recipe, which is superb with paella-style rice dishes.

1 Prepare the alioli first. Crush the garlic and place it in a blender or food processor with seasoning. Add the egg and lemon juice and process until pale and frothy. With the motor running, gradually drizzle in the oil, until the mixture thickens. Season to taste and transfer to a serving dish. Cover and leave at room temperature.
2 Cut the artichokes and aubergines in half from tip to stalk. Brush with a little oil and cook on the barbecue for 10 minutes, turning occasionally until starting to blacken. Garnish with lemon wedges and serve with the alioli.

Variation
NEW POTATOES WITH ALIOLI For a quick starter, simmer baby new potatoes in water until almost tender, then drain and place on the barbecue for 5 minutes, until golden. Serve the potatoes on wooden cocktail sticks, with the alioli.

Cook's Tip
Have the egg at room temperature, not chilled, or the alioli may curdle. If it does, start again with another egg, then drizzle in the curdled mixture as the motor is running. Add another 155 ml (5 fl oz) olive oil or the sauce will be 'eggy'.

If making the alioli in advance, cover it or place it in an airtight container and chill. Then bring to room temperature for about 30 minutes before serving.

GARLIC MUSHROOMS

Time to prepare: 10 minutes
Time to cook: 15 minutes

Serves 4

8 large flat open mushrooms

Garlic Butter

60 g (2 oz) butter
2 garlic cloves
3 tablespoons chopped fresh flat-leaf parsley
a few drops Tabasco sauce
1 teaspoon grated lemon rind
salt and freshly ground black pepper

Large flat open mushrooms, sometimes known as field mushrooms, are an excellent choice for cooking on the barbecue and this incredibly simple treatment takes some beating. I use chopped garlic in my garlic butter as I like the chewy little pieces that really burst with flavour when you bite them; crush or mash the garlic if you are not such a fan. Serve lots of crusty bread to mop up the garlic butter.

1 Wipe the mushrooms and remove their stalks. To prepare the garlic butter, cream the butter until softened. Finely chop the garlic and beat it into the butter with the parsley, Tabasco, lemon rind and plenty of seasoning.
2 Place the mushrooms on the barbecue, stalk sides down, and cook for about 10 minutes, until blackened.
3 Turn the mushrooms over and fill the tops with garlic butter. Cook for a further 5 minutes, until the butter has melted and the mushrooms are tender. Serve immediately.

BARBECUED ASPARAGUS WITH PARMESAN AND PINE NUT DRESSING

Time to prepare: 15 minutes
Time to cook: 10 minutes

Serves 4

500 g (1 lb) fresh young asparagus
a little olive oil, for brushing

Parmesan and Pine Nut Dressing
30 g (1 oz) pine nuts
1 garlic clove, peeled
a little rock salt
4 tablespoons freshly grated Parmesan cheese
2 tablespoons chopped fresh flat-leaf parsley
4 tablespoons extra virgin olive oil
freshly ground black pepper

We are used to asparagus served with rich buttery sauces, but its distinctive flavour is also excellent with a good olive-oil-based dressing. Barbecuing the asparagus spears also accentuates their flavour – a tip I learned from inspirational food writer, Nigel Slater. This recipe is ideal for the first barbecue of the summer, if we are lucky enough to have a few hot days in late May or June to coincide with the English asparagus season.

1 Trim the asparagus, brush with a little olive oil and set aside while making the dressing.
2 Place the pine nuts and garlic in a mortar with a sprinkling of rock salt. Use a pestle to pound the ingredients together until you have a smooth paste. Then work in the Parmesan and parsley. Gradually add the olive oil a little at a time until you have a thick dressing. Check the seasoning, adding freshly ground black pepper to taste.
3 Cook the asparagus on the barbecue for about 10 minutes, until tender. Turn the spears regularly and brush with a little more oil if they become too dry. Serve with the dressing and crusty Italian bread to mop up any juices.

Cook's Tip
For a fabulous pasta dish, cut up the asparagus and toss it with the dressing, then serve with hot freshly cooked tagliarini pasta.

HERB AND GINGER CHEESE WITH BARBECUED VEGETABLES

Time to prepare: 15 minutes
Time to cook: 10 to 15 minutes

Serves 4

125 g (4 oz) baby carrots
125 g (4 oz) baby fennel
125 g (4 oz) baby leeks
125 g (4 oz) even-sized chestnut mushrooms
2 tablespoons extra virgin olive oil

Herb and Ginger Cheese
250 g (8 oz) soft cheese, such as medium-fat soft cheese, ricotta or cream cheese
125 g (4 oz) Greek-style yogurt
2.5 cm (1 in) piece preserved stem ginger
4 tablespoons chopped mixed fresh herbs, such as chives, parsley, tarragon and basil
salt and freshly ground black pepper
flat-leaf parsley, to garnish

1 First prepare the herb and ginger cheese. Place the cheese in a bowl with the Greek yogurt and mix well. Finely chop the ginger and stir in with the herbs and seasoning to taste.
2 Trim and scrub the carrots. Wash the fennel and leeks thoroughly. Wipe the mushrooms and remove their stems. Blanch the carrots in a pan of boiling water for 5 minutes, adding the fennel and leeks for the last 2 minutes. Drain the vegetables, plunge them into cold water to prevent them from cooking further and drain well.
3 Brush all the vegetables with olive oil. Place on the barbecue and cook for about 10 minutes, turning regularly, until golden and tender.
4 Transfer the vegetables to plates. Garnish with flat-leaf parsley and serve at once, with spoonfuls of the cheese.

Variation
FRESH YOGURT CHEESE Home-made fresh yogurt cheese makes the most wonderful accompaniment for barbecued baby vegetables but it has to be prepared 24 hours in advance. The process is so easy that I feel it deserves to be mentioned as the final cooking is quick and in the spirit of the book.

Prepare the cheese at least 24 hours ahead. Place 250 g (8 oz) Greek-style yogurt in a bowl with 250 g (8 oz) crème fraîche and mix well. Finely chop the ginger and stir in with the herbs as in the main recipe; season to taste. Place in the centre of a large square of double-thickness butter muslin or clean J-cloth. Draw the corners together to form a bag and tie securely with string to form a pudding shape.

Hang the bag in a cool place and place a bowl underneath to catch any liquid. Leave for 24 hours until the cheese is firm. Chill until ready to serve.

GRILLED BEETROOT WITH PEAR AND ROCKET SALAD

Time to prepare: 10 minutes
Time to cook: 10 minutes

Serves 4

8 small cooked beetroots
2 tablespoons extra virgin olive oil
1 bag rocket leaves
1 ripe pear
juice of ½ lemon

Walnut Cream Sauce

6 tablespoons créme fraîche
1 teaspoon Dijon mustard
30 g (1 oz) walnut pieces, finely chopped
2 tablespoons chopped mixed fresh herbs such as dill, chives, parsley and tarragon
salt and freshly ground black pepper

This is a version of a recipe that has already appeared in a previous book but which works so well on the barbecue that I had to include it here. Baby beetroots are sweet and tender when cooked this way. For the best flavour, make sure you buy beetroot that is packed without vinegar or acid.

1 Prepare the sauce first: mix together the créme fraîche, mustard, chopped walnuts and herbs. Season to taste, cover and chill.
2 Brush each beetroot with a little olive oil. Season and cook on the barbecue, turning occasionally, for about 10 minutes.
3 To serve, arrange a bed of rocket leaves on each of four plates. Quarter and core the pear and cut it into thin slices, then toss in the lemon juice and scatter over the rocket.
4 Cut the beetroots into quarters and arrange on the leaves. Spoon over a little sauce and serve immediately.

Cook's Tip

The beetroot may be wrapped in foil and cooked directly on the coals. Tuck the foil parcels in among the coals while the main course cooks on top and you will not use up precious grid space. Place each beetroot on a square of foil and drizzle with a little olive oil. Season and wrap each beetroot in foil to make a small parcel. Tuck the parcels in among the hot coals and cook for 15 to 20 minutes.

ROASTED AUBERGINE AND CORIANDER PATE

Time to prepare: 15 minutes
Time to cook: 15 minutes

Serves 4

2 medium aubergines
3 tablespoons olive oil
½ teaspoon cayenne pepper
1 teaspoon ground cumin
1 garlic clove, crushed
juice of ½ lemon
2 tablespoons chopped fresh coriander
salt and freshly ground black pepper
fingers of toasted brioche or pitta bread, to serve

Writing cookery books is an excellent excuse, if one is needed, to eat out as much as possible as it is vital for inspiration, or so I keep telling myself. This pâté is based on an aubergine dip which I was served as part of a range of intricate '*amuse gueule*' at a Michelin-starred restaurant in Kent. It came as a topping for tiny little morsels of brioche toast and had a garnish of diced tomato, roasted red pepper and chervil leaves. I have transformed it into a pâté to be served with brioche toast, based on the Middle Eastern classic *baba ganoush* also called 'poor man's caviar'. Barbecuing aubergine intensifies its wonderful smoky flavour until it is practically irresistible.

1 Cut the aubergines in half from stalk to base. Slash the cut surfaces with a knife into a criss-cross pattern. Mix half the olive oil with the cayenne and cumin, and brush over the cut surfaces of the aubergine.
2 Place the aubergine halves on the barbecue, cut sides up, and barbecue for 10 to 15 minutes until blackened. Turn and cook the other sides until golden – the flesh must be really tender. Scrape the flesh out of the blackened skins into a blender or mixing bowl. Process or mash until smooth.
3 Add the garlic, then gradually add the lemon juice with the motor running or while beating continuously by hand. Mix in the coriander. Season to taste and serve with the toasted brioche or pitta bread.

Cook's Tip
This recipe works well with fresh mint instead of coriander and it can be transformed into a wonderful salad by adding 2 seeded and chopped tomatoes, 1 chopped red pepper and a finely chopped red onion.

Kebabs

The idea that vegetarians always take second best at barbecues is certainly addressed in this chapter, which really shows the versatility of vegetables for creating a whole range of colourful, attractive kebabs. These skewered foods are all far more interesting to cook and eat than dull-tasting chunks of meat!

CHERRY TOMATO AND GARLIC KEBABS

Time to prepare: 5 minutes
Time to cook: 10 minutes

Serves 4

2 yellow peppers
400 g (14 oz) can artichoke hearts
12 very large garlic cloves
20 cherry tomatoes
16 fresh bay leaves
2 tablespoons olive oil
1 teaspoon dried oregano
1 tablespoon balsamic vinegar
salt and freshly ground black pepper

These kebabs are definitely for garlic lovers – if you are not a garlic fan you can use shallots instead. Look out for giant garlic in the supermarket; the harsh flavour of the cloves is softened and the cloves grow sweet as they cook. Serve these kebabs with an olive oil bread, such as focaccia or ciabatta, to mop up all the juices.

1 Halve and seed the peppers, then cut them into 2.5 cm (1 inch) squares. Drain the artichoke hearts and peel the garlic cloves.
2 Thread the peppers, artichoke hearts and garlic onto eight skewers with the cherry tomatoes and bay leaves. Mix the oil, oregano, balsamic vinegar and seasoning.
3 Place the kebabs on the barbecue and brush with the oil mixture. Cook for 8 to 10 minutes, turning and brushing with the oil mixture, until the skins of the tomatoes are charred. Serve freshly grilled.

Cook's Tips
Italian cooks always prefer using dried oregano to the fresh herb as it has a stronger, more distinctive, flavour; however, if you find it overpowering, substitute fresh oregano or use thyme or basil instead. If you can find them, substitute 8 young baby artichokes (the type that are eaten whole) for the can of artichoke hearts.

Serving Cherry Tomato and Garlic Kebabs with Strips of Garlic Herb Pizza Bread, see page 73

THAI SPICED MUSHROOM KEBABS

Time to prepare: 15 minutes
Time to marinate: 30 minutes
Time to cook: 5 minutes

Serves 4

375 g (12 oz) shiitake mushrooms
3 garlic cloves
1 teaspoon Thai red curry paste
2 tablespoons chopped fresh coriander
1 teaspoon grated fresh root ginger
2 tablespoons light soy sauce
1 tablespoon dark muscovado sugar

Chilli and Cucumber Pickle
10 cm (4 inch) piece cucumber
1/2 red onion
1–2 small red chillies
4 tablespoons rice vinegar or cider vinegar
1 tablespoon caster sugar
1 teaspoon salt

Shiitake mushrooms work best for these kebabs but if you cannot find them, use medium chestnut mushrooms instead. Make sure that the red curry paste you buy does not contain shrimp paste.

1 Cut the mushrooms into quarters and place in a bowl. Crush the garlic and mix together with the red curry paste, coriander, ginger, soy sauce and muscovado sugar. Pour the paste over the mushrooms, mix lightly and leave to marinate for 30 minutes.
2 To make the pickle, cut the cucumber in quarters lengthways. Thinly slice the lengths to give small wedge-shaped pieces of cucumber. Finely chop the red onion, seed and chop the chillies. Stir the vinegar, caster sugar and salt together in a small bowl until the sugar dissolves. Add the cucumber, onion and chillies and mix well.
3 Remove the mushrooms from the marinade and thread them onto bamboo skewers. Place on the barbecue and cook for 5 minutes, turning frequently. Serve at once with the cucumber pickle.

NEW POTATO AND SHALLOT KEBABS WITH CORIANDER PESTO

Time to prepare: 15 minutes
Time to marinate: 30 minutes
Time to cook: 15 to 20 minutes

Serves 4

500 g (1 lb) even-sized new potatoes
250 g (8 oz) shallots
1 garlic clove
½ green chilli
3 tablespoons olive oil
1 teaspoon grated lemon rind
salt and freshly ground black pepper

Coriander Pesto
30 g (1 oz) fresh coriander leaves
15 g (½ oz) walnut pieces
1 garlic clove
3 tablespoons olive oil

Potatoes and onions are perfect partners and they feature together in dishes around the world. Shallots are used here but the kebabs can also be made with quartered onions, such as red onions, which have a mild flavour. Use metal skewers which conduct the heat to the centre of the potatoes and ensure they cook through properly. I suggest marinating the vegetables for 30 minutes, but this can be reduced to 15 minutes if you are in a hurry.

1 Scrub the potatoes and cook in a medium saucepan of simmering water for 5 minutes. Add the shallots for the last 2 minutes. Drain the vegetables and peel the shallots.
2 Crush the garlic, seed and chop the chilli, and mix both with the oil, lemon rind and plenty of seasoning. Pour the mixture over the vegetables and leave them to marinate for 30 minutes.
3 Meanwhile, place the coriander leaves in a food processor or blender with the walnut pieces and garlic. Process until smooth, then gradually drizzle in the oil with the motor running to form a thick paste. Season to taste and transfer to a serving dish.
4 Thread the potatoes and shallots onto skewers. Place the kebabs on the barbecue and cook for 15 to 20 minutes, until tender, turning occasionally and brushing with the marinade. Serve the coriander pesto with the kebabs.

Cook's Tip
You can serve bought pesto instead of the coriander pesto if you are short of time.

SWEETCORN AND RED ONION KEBABS WITH SESAME OIL

Time to prepare: 15 minutes
Time to marinate: 30 minutes
Time to cook: about 15 minutes

Serves 4

4 corn cobs
4 medium red onions
185 g (6 oz) broccoli
1 garlic clove
2 tablespoons olive oil
2 tablespoons teriyaki sauce
1 tablespoon sesame oil
juice of 1 lemon
salt and freshly ground black pepper
freshly cooked jasmine rice, to serve

Sesame oil is a full-flavoured, fragrant ingredient to keep in every larder and it smells wonderful when brushed over vegetables as they grill. Here, sesame oil adds its distinctive character to a simple marinade based on bought teriyaki sauce. I suggest marinating the kebabs for 30 minutes, but this can be reduced to 15 minutes if you are in a hurry.

1 Remove the husks from the corn and cut each cob into four. Cut the onions into quarters and break the broccoli into florets. Cook the corn and broccoli in a large saucepan of boiling water for 2 minutes. Drain and refresh under cold water, then drain well.
2 Thread the corn, onion quarters and broccoli onto skewers. Place the kebabs in a shallow dish.
3 Crush the garlic in a small bowl. Mix in the olive oil, teriyaki sauce, sesame oil, lemon juice and seasoning. Pour over the kebabs and leave to marinate for 30 minutes, turning occasionally.
4 Cook the kebabs on the barbecue for about 15 minutes, turning regularly and brushing with the remaining marinade, until the vegetables are beginning to char. Serve with freshly cooked jasmine rice.

SPICED ROOT VEGETABLE KEBABS WITH RHUBARB AND ONION CHUTNEY

Time to prepare: 10 minutes
Time to cook: 30 minutes

Serves 4

500 g (1 lb) mixed root vegetables, such as carrots, parsnips, new potatoes and swede
8 large spring onion bulbs
1 tablespoon biryani paste or hot curry paste
1 teaspoon tomato purée
2 tablespoons sunflower oil
salt and freshly ground black pepper

Onion and Rhubarb Chutney
250 g (8 oz) rhubarb
1 medium onion
1 tablespoon sunflower oil
1 teaspoon white mustard seeds
½ teaspoon ground turmeric
30 g (1 oz) raisins
125 ml (4 fl oz) cider vinegar
90 g (3 oz) demerara sugar

A good-quality biryani paste is a useful store-cupboard ingredient as it can be brushed over grilling vegetables as in this recipe. If you prefer a spicier flavour, use a hot curry paste, but make sure that it is a good-quality variety. Serve these kebabs on a bed of pilau rice or with the Couscous Salad with Barbecued Vegetables, see page 59.

1 First prepare the chutney: slice the rhubarb and onion. Heat the oil in a small pan and add the onion. Cook for 3 minutes, until softened, then add the rhubarb, mustard seeds, turmeric, raisins, cider vinegar and sugar. Bring to the boil and simmer for 20 minutes without a lid, until thickened. Transfer to a serving dish.
2 Meanwhile, cut the root vegetables into 5 cm (2 inch) cubes and blanch them in boiling water for 5 minutes. Drain and refresh under cold water.
3 Thread the vegetables onto metal skewers with the spring onion bulbs. Mix the biryani or hot curry paste with the tomato purée and sunflower oil. Add seasoning and brush the paste over the vegetables.
4 Cook the vegetable kebabs on the barbecue for 10 minutes, turning regularly, until tender and golden. Serve at once, with the chutney.

Cook's Tip

If you do not have time to make the chutney, serve a good mango chutney instead.

PUMPKIN AND CELERIAC KEBABS WITH CINNAMON LIME BUTTER

Time to prepare: 10 minutes
Time to cook: about 20 minutes

Serves 4

375 g (12 oz) pumpkin flesh
375 g (12 oz) celeriac
60 g (2 oz) butter
1 tablespoon clear honey
grated rind and juice of 1 lime
1/2 teaspoon ground cinnamon
salt and freshly ground black pepper

Halloween always tests my creativity for pumpkin recipes as the house overflows with pumpkin lanterns and I cannot bring myself to throw away the flesh that is scooped out. I first discovered how well pumpkin and celeriac go together when I paired them up in a soup; the sweetness of the pumpkin is balanced by the cleaner flavour of the celeriac.

1 Peel the pumpkin and celeriac and cut the flesh into 5 cm (2 inch) cubes. Bring a medium saucepan of water to the boil, add the vegetables and cook for 5 minutes. Then drain and refresh the vegetables under cold water.
2 Melt the butter with the honey in a small saucepan and stir in the lime rind and juice, cinnamon and seasoning.
3 Thread the vegetable cubes onto skewers and brush with the butter. Cook on the barbecue for 15 minutes, turning regularly and brushing with the butter, until the vegetables are golden and tender.

Cook's Tip
Any combination of root vegetables will work well on these kebabs: for example, try sweet potato, swede and parsnip.

VEGETABLE SATAY

Time to prepare: 15 minutes
Time to marinate: 1 hour
Time to cook: 15 minutes

Serves 4

2 x 250 g (8 oz) blocks firm tofu
1 medium carrot
1 medium courgette

Marinade
2 teaspoons ground cumin
4 tablespoons dark soy sauce
2 tablespoons red wine vinegar
2 tablespoons light muscovado sugar
Peanut Satay Sauce, see page 81, to serve

Garnish
lime wedges
flat-leaf parsley

For classic satay, the chosen meat is cut into thin pieces then marinated with sugar, spices and a little soy sauce so that it cooks through quickly without drying out. Here, strips of courgette and carrot are wrapped around cubes of marinated tofu and served with the same peanut sauce to echo the classic flavourings. I suggest marinating the tofu for 1 hour but this can be reduced to 30 minutes if you are in a hurry.

1 Cut the tofu into 4 cm (1½ inch) cubes and place in a shallow dish. For the marinade, mix the cumin, soy sauce, vinegar and sugar and pour this over the tofu. Cover and chill for 1 hour, turning occasionally.
2 Just before serving, trim off the ends of the carrot and courgette. Using a vegetable peeler, pare lengths off the vegetables making thin ribbons. Wrap a ribbon of courgette or carrot around each cube of tofu and thread onto metal skewers.
3 Place the kebabs on the barbecue and cook for 5 to 10 minutes, turning occasionally until the vegetables are golden. Garnish with lime wedges and flat-leaf parsley and serve with the peanut sauce.

AUBERGINE AND SMOKED TOFU KEBABS

Time to prepare: 10 minutes
Time to cook: about 15 minutes

Serves 3 to 4

1 large aubergine
250 g (8 oz) block smoked tofu
12 fresh kaffir lime leaves or fresh bay leaves
1 teaspoon cumin seeds
2 dried chillies
3 tablespoons sunflower oil

One of pleasures I enjoyed most while I was editor of *Vegetarian Good Food* magazine, was tasting the recipes that were cooked in the test kitchen before they appeared in the magazine. The inspiration for this recipe is now a dish that has become a family favourite. Its creator, food writer Rosemary Stark, served the kebabs on a bed of lentil dhal flavoured with coconut and lime – a sublime combination of flavour, colour and texture. For this barbecue version, I suggest serving the kebabs with Rocket with Char-grilled Red Pepper and Caramelised Shallots, see page 62, or with Peanut Satay Sauce, see page 81.

1 Cut the aubergine and tofu into 5 cm (2 inch) cubes and thread them onto bamboo skewers, alternating with the kaffir lime leaves or bay leaves.
2 Heat a small frying pan, then add the cumin seeds and chillies. Fry for 30 seconds, pour in the oil and remove from the heat; leave to infuse for 5 minutes.
3 Brush the kebabs with the infused oil and cook on the barbecue for 15 minutes, turning regularly, until the tofu is crisp and golden and the aubergine has charred. Serve the kebabs at once.

Cook's Tip
If you cannot find smoked tofu, try marinating plain tofu instead. Mix equal parts of soy sauce and dry sherry, and pour this over the cubed tofu. Marinate for 1 hour, then drain and pat the tofu dry on absorbent kitchen paper. Use as above.

AUBERGINE AND MOZZARELLA ROLLS

Time to prepare: 15 minutes
Time to cook: 9 to 11 minutes

Serves 4

1 large or 2 small aubergines
3 tablespoons olive oil
2 tablespoons sun-dried tomato paste
375 g (12 oz) mozzarella cheese
24 large fresh basil leaves
salt and freshly ground black pepper

Melanzane Parmigiano is one of my favourite supper dishes: it consists of layers of fried aubergine, tomato and mozzarella cooked together until the flavours are well combined. These kebabs are based on the same ingredients, with the additional flavour of barbecue cooking. Watch the kebabs carefully during cooking as the mozzarella can get quite runny and messy, so be ready to scrape off any that threatens to fall through the grid onto the coals.

1 Cut the aubergine lengthways into eight 5mm (¼ in) thin slices and then cut each slice across in half again. Brush the slices with half the olive oil and cook briefly on the barbecue for 2 minutes on each side until just charred. Allow to cool for 5 minutes.
2 Spread each slice of aubergine with a little sun-dried tomato paste. Cut the mozzarella into 16 strips and place one in the centre of a slice of aubergine. Add a basil leaf and season well. Wrap up to form a small parcel and thread onto four skewers.
3 Brush the skewered rolls with the remaining oil and place on the barbecue. Cook for 5 to 7 minutes, turning occasionally, until the aubergine is browned.

Cheese, pulses and tofu

This chapter is a fairly eclectic mix of recipes that I have grouped together for their convenience. Based on cheese, pulses or tofu, they all make excellent main dishes for barbecues and I find them especially popular with children.

MOZZARELLA-STUFFED RADICCHIO WITH BLACK OLIVE RELISH

Time to prepare: 15 minutes
Time to cook: 7 to 8 minutes

Serves 4

4 medium heads radicchio
125 g (4 oz) buffalo mozzarella cheese
2 tablespoons extra virgin olive oil

Black Olive Relish
60 g (2 oz) black olives, stoned
4 sun-dried tomatoes in oil, drained
1 shallot
4 tablespoons chopped fresh flat-leaf parsley
1 teaspoon capers
2 tablespoons extra virgin olive oil
salt and freshly ground black pepper

Radicchio has quite a pungent, bitter flavour which mellows and becomes sweeter when cooked on the barbecue. Use the round variety for this recipe as it is easier to stuff. It is actually a type of chicory, so if you cannot buy radicchio, chicory is a reasonable alternative.

1 Drop the radicchio heads into a saucepan of boiling water and cook for 2 to 3 minutes. Remove from the water with a draining spoon and pat dry on absorbent kitchen paper. Brush with olive oil and leave to stand for 10 minutes.
2 Meanwhile, make the relish. Finely chop the olives, sun-dried tomatoes and shallot. Mix with the parsley, capers, oil and plenty of seasoning. Place the relish in a serving dish and set aside.
3 Cut the mozzarella into four cubes. Carefully open the leaves of the radicchio and ease a cube of mozzarella into the centre of each head. Fold over the leaves to enclose the cheese securely and place on the barbecue. Cook the radicchio for about 5 minutes, until charred and wilted.

Cutting into freshly cooked Mozzarella-stuffed Radicchio served with Black Olive Relish

PEKING-STYLE MARINATED TOFU STEAKS WITH PLUM SAUCE

Time to prepare: 10 minutes
Time to marinate: 24 hours
Time to cook: 7 to 11 minutes

Serves 4

315 g (10 oz) block tofu
1 teaspoon Szechwan peppercorns
1 teaspoon salt
1 teaspoon grated fresh root ginger
1/2 teaspoon five spice powder
1 tablespoon honey
1 tablespoon rice wine or dry sherry
1 teaspoon sesame oil
1 tablespoon light soy sauce

To Serve
6 spring onions
10 cm (4 inch) piece cucumber
ready-made plum sauce
8 Chinese pancakes

I love the flavours associated with Peking duck – the pleasing balance of sweet and sour, the crisp and soft textures provided by the vegetables, pancakes and sweet plum sauce. Tofu takes up the flavours of the classic marinade and crisps well on the barbecue, so vegetarians do not have to miss this meal usually centred on duck. Chinese pancakes are readily available in supermarkets.

1 Cut the tofu into four strips and place them in a shallow dish. Place the peppercorns in a dry frying pan and fry them for 1 minute.
2 Use a pestle to grind the Szechwan peppercorns in a mortar. Then mix in the salt, ginger, five spice powder, honey, rice wine or dry sherry, sesame oil and light soy sauce. Pour over the tofu, cover and leave to marinate for 24 hours, if possible.
3 Before cooking the tofu, cut the spring onions into short thin strips. Peel and cut the cucumber into similar-sized strips.
4 Remove the tofu from the marinade and cook on the barbecue for 3 to 4 minutes on each side, brushing with the marinade, until golden. Remove from the barbecue and cut into thin strips.
5 To serve, brush a little plum sauce onto each pancake and top with tofu, spring onion and cucumber. Roll and serve.

Cook's Tip

Szechwan peppercorns are available in good supermarkets and specialist Chinese shops. If you cannot find them, use black peppercorns instead.

BARBECUED HALLOUMI STRIPS

Time to prepare: 15 minutes
Time to marinate: overnight
Time to cook: 10 to 12 minutes

Serves 4

500 g (1 lb) halloumi cheese
1 lime
2.5 cm (1 inch) piece fresh root ginger
1 garlic clove
3 tablespoons chopped fresh coriander
3 tablespoons olive oil
salt and freshly ground black pepper

Halloumi is a Greek cheese that has a firm, pliable texture, rather like mozzarella, but it does not melt and go runny in the same way, so it holds it shape well when cooked, which makes it ideal for barbecuing. Though rather bland on its own, halloumi absorbs flavours well and it is excellent with a well-flavoured marinade, such as the lime and ginger one used here. Rocket with Char-grilled Red Pepper and Caramelised Shallots, page 62, goes well with the barbecued halloumi.

1 Cut the cheese into eight thin slices and arrange them in a single layer in a large shallow dish.
2 Finely grate the lime rind and squeeze the juice. Peel and finely chop the ginger and garlic. Mix the lime rind and juice, ginger, garlic, coriander, olive oil and seasoning. Pour the mixture over the halloumi, cover and leave to marinate for at least 12 hours.
3 When you are ready to cook the halloumi, remove the cheese from the marinade and place it on the barbecue. Cook for 10 to 12 minutes, turning regularly and brushing with the marinade, until the cheese starts to turn golden.

SPICED RED BEAN KOFTAS WITH MANGO AND COCONUT CHUTNEY

Time to prepare: 20 minutes
Time to cook: 9 to 11 minutes

Serves 4

400 g (14 oz) can red kidney beans, well drained
1 teaspoon garam masala
½ teaspoon ground coriander
¼ teaspoon ground cumin
½ small onion
1 garlic clove
½ green chilli
2 tablespoons olive oil
salt and freshly ground black pepper

Mango and Coconut Chutney
1 green chilli
4 tablespoons unsweetened desiccated coconut
4 tablespoons chopped fresh coriander
3.5 cm (1½ inch) piece fresh root ginger
1 garlic clove
1 tablespoon fresh lime juice
1 small ripe mango
flat-leaf parsley, to garnish

Koftas are spiced Indian patties made from meat, dhall or vegetables. Made here with red kidney beans, they are ideal candidates for cooking on the barbecue. Serve them with their hot mango and coconut chutney, a green salad and plenty of Indian bread, such as naan, or soft flour tortillas.

1 Place the beans in a bowl with the garam masala, coriander and cumin. Finely chop the onion and garlic. Seed and finely chop the chilli.
2 Heat the oil in a small frying pan and cook the onion, garlic and chilli for 3 minutes until the onion is softened. Add to the beans with plenty of seasoning and mash the ingredients together thoroughly. Shape the mixture into 16 small balls and thread them onto small wooden skewers.
3 Make the chutney: seed and roughly chop the chilli, then place it in a blender or food processor. Add the coconut, coriander, ginger, garlic and lime juice. Blend the ingredients to a smooth paste, then add salt to taste.
4 Peel the mango. Cut the flesh away from the stone. Cut it into small cubes and place in a serving bowl. Pour the coconut mixture over the mango and mix well.
5 To cook the kofta, place the skewers on the hot barbecue and cook for 6 to 8 minutes, turning regularly. Serve at once, garnished with flat-leaf parsley and accompanied by the mango and coconut chutney and warmed naan bread.

MUSHROOM CARROT BURGERS WITH FETA CHEESE

Time to prepare: 15 minutes
Time to cook: 18 to 23 minutes

Serves 4

1 medium onion
2 garlic cloves
375 g (12 oz) chestnut mushrooms
375 g (12 oz) carrots
2 tablespoons olive oil
125 g (4 oz) fresh wholemeal breadcrumbs
90 g (3 oz) feta cheese
3 tablespoons chopped fresh basil
2 tablespoons chopped fresh thyme
salt and freshly ground black pepper
2 eggs

This burger mixture is based on a recipe from the *Moosewood Cookbook* by Mollie Katzen. I bought this wonderful cookery book on my first trip to New York over twenty years ago, when vegetarianism was in its relative infancy and seemed to be based entirely on lentils and nuts. The *Moosewood Cookbook*, with its imaginative salads, wonderful breads and really inspiring alternatives to meat-based dishes, became my first veggie bible and I still turn to it for ideas or to revisit old favourites.

1 Finely chop the onion and garlic. Wipe and finely chop the mushrooms. Grate the carrots. You will save time if you use a food processor for all this preparation.
2 Heat the oil in a large frying pan and add the onion and garlic. Cook for 3 minutes, until softened, but not browned. Stir in the mushrooms and cook for a further 5 minutes.
3 Place the grated carrots in a large mixing bowl with the breadcrumbs. Crumble the feta and add it to the bowl with the mushroom mixture. Thoroughly mix in the basil, thyme and plenty of seasoning. Beat the eggs and stir them in to bind the mixture.
4 Divide the mixture into four and pat into flat burgers.
5 Use a fish slice or spatula to place the burgers on the barbecue and cook for 10 to 15 minutes, turning regularly, until the burgers are golden. Serve with a leafy green salad.

Cook's Tip
To freeze the burgers, place them on a baking sheet lined with baking parchment and open freeze for several hours. Wrap tightly and label. Freeze for up to a month. To use, thaw thoroughly, then cook as above.

BARBECUED GOATS' CHEESE AND HERB FOCACCIA

Time to prepare: 10 minutes
Time to cook: about 4 minutes

Serves 1

- **1 small garlic clove**
- **1 spring onion**
- **60 g (2 oz) fresh goats' cheese**
- **2 tablespoons chopped fresh herbs, such as basil, chives, marjoram and flat-leaf parsley**
- **salt and freshly ground black pepper**
- **1 piece focaccia, about 10 cm (5 inch) square**
- **1 tablespoon extra virgin olive oil**
- **1 ripe plum tomato**

This is actually a glorified toasted sandwich, cooked on the barbecue, but it really is wonderful. Now that it is so easy to find excellent-quality speciality breads in local supermarkets, this recipe is simple to prepare. I use a plain rosemary focaccia, but any flavoured flat Italian bread will work well, including varieties flavoured with ingredients ranging from mozzarella and garlic to sun-dried tomato and herbs. Choose a soft fresh goats' cheese for this recipe which makes a good informal starter when cut into strips and eaten with the fingers rather than knives and forks.

1 Finely chop the garlic and spring onion. Place in a bowl with the cheese and herbs. Season well and mash the ingredients until thoroughly mixed.
2 Split the focaccia in half horizontally and drizzle olive oil over each cut side. Slice the tomato. Spread the cheese mixture on one piece of bread. Top with tomato slices and the other piece of focaccia. Press the top down firmly to flatten the sandwich.
3 Cook the focaccia sandwich on the barbecue for about 2 minutes on each side, until the cheese melts and the bread is golden and crisp. Take care not to let the bread burn before the inside is heated through. Serve immediately.

Cook's Tip
You can try all kinds of fillings for the focaccia, but a cheese one that melts works best. For example, try mozzarella and barbecued peppers or feta with black olive paste and artichokes.

CHAR-GRILLED PEPPERS AND NEW POTATOES WITH MELTING FONTINA

Time to prepare: 20 minutes
Time to cook: 20 to 25 minutes

Serves 4

2 large red peppers
2 large orange peppers
750 g (1½ lb) small new potatoes
3 tablespoons olive oil
1 tablespoon chopped fresh thyme
1 tablespoon grainy mustard
185 g (6 oz) Fontina cheese
salt and freshly ground black pepper

The inspiration for this recipe comes straight from recent skiing holidays spent in the Haute Savoie region of France. Melted cheese dishes served with large amounts of carbohydrate in the form of bread and potatoes are the perfect comfort food for bruised and tired skiers – at least, I am usually in such condition and need little excuse for a splendid meal washed down with large quantities of mulled wine. Though this seems a somewhat heavy dish for warm summer nights, the addition of sweet peppers balances the richness and the proportion of cheese is reduced to avoid indigestion! The cheese must be melted at the last minute and spooned over the vegetables; I suggest using a heavy iron baking sheet or grill pan for this purpose.

1 Halve and seed the peppers, then cut them into quarters. Cook the potatoes in a medium saucepan of boiling water for 8 minutes, then drain.
2 While the potatoes are cooking, mix the olive oil with the thyme and mustard. Pour this mixture over the drained potatoes and leave to stand for 5 minutes.
3 Thread the potatoes and peppers onto metal skewers. Brush with any remaining oil and cook on the barbecue for 10 to 15 minutes, until the potatoes are golden and the peppers charred. Slide the peppers and potatoes off the skewers into a shallow serving dish and keep warm.
4 Thinly slice the cheese. Heat a baking sheet or large flat pan on the barbecue and lay the cheese on the hot surface. As the cheese melts, scrape it off the sheet or pan and onto the vegetables. Serve at once with crusty bread or Garlic Herb Pizza Bread, see page 73, to mop up the juices.

Cook's Tip
If you cannot find Fontina, use a cheese that melts easily, becoming quite runny so that it can be scraped off the sheet or pan; for example, try Emmental, Gruyère or Gouda.

FALAFEL IN PITTA POCKETS WITH TAHINI MINT DRESSING

Time to prepare: 20 minutes
Time to cook: 6 minutes

Serves 4

2 x 400 g (14 oz) cans chick peas, drained
60 g (2 oz) fresh wholemeal breadcrumbs
2 garlic cloves
1 small red chilli, optional
1 stick celery
2 spring onions
2 teaspoons ground cumin
2 teaspoons ground coriander
1/4 teaspoon turmeric
2 tablespoons olive oil
salt and freshly ground black pepper

Yogurt Dressing

6 tablespoon Greek-style yogurt
2 tablespoons chopped fresh mint
1 tablespoon tahini paste

To Serve

4 pieces pitta bread
tomato and cucumber salad

These spicy little patties cook well on the barbecue and they can be served as part of a satisfying mezze-style spread, with the Couscous Salad with Barbecued Vegetables, see page 59, Roasted Aubergine and Coriander Pâté, see page 21, and Vine-leaf Parcels with Feta and Marinated Olives, opposite.

1 Place the chick peas in a food processor with the breadcrumbs.
2 Crush the garlic and seed and chop the chilli if using. Cut the celery and spring onions into short lengths and add all these ingredients to the chick peas. Process until smooth, then add the cumin, coriander, turmeric, olive oil and salt to taste. Process again to mix.
3 Shape the mixture into eight small patties and chill them for 15 minutes.
4 Meanwhile prepare the yogurt dressing. Gradually stir the yogurt and mint into the tahini. Season to taste and set aside.
5 Place the falafel on the hot barbecue and cook for 3 minutes on each side, until golden. Warm the pitta bread on the back or to one side of the barbecue, cut in half and open into pockets.
6 To serve, slip a couple of falafel into each pitta pocket and drizzle in a little of the dressing. Serve immediately, with a tomato and cucumber salad.

Cook's Tip
Tahini tends to separate when stored, so stir it well in the jar to combine the thick paste with the oil on the surface before measuring out the required quantity.

VINE-LEAF PARCELS WITH FETA AND MARINATED OLIVES

Time to prepare: 15 minutes
Time to cook: 4 to 6 minutes

Serves 4

150 g (6 oz) feta cheese
60 g (2 oz) marinated black olives, stoned
1 garlic clove
1 teaspoon capers
2 tablespoons chopped fresh marjoram or basil
freshly ground black pepper
16 large vine leaves in brine, drained
2 tablespoons olive oil

I find that vine leaves can be rather slimy and chewy, and I often struggle with them when they are served as part of a mezze. However, their acidic flavour and tough texture make them an ideal wrapping for food cooked on the barbecue. With a full-flavoured cheese filling, they are transformed into a smoky treat. They are very salty, so do not add any extra salt to the filling. I like to use marinated olives for this recipe and usually buy the ones in olive oil flavoured with garlic and lemon.

1 Cut the feta into small cubes and coarsely chop the olives, garlic and capers. Mix with the marjoram or basil and plenty of freshly ground black pepper.
2 Rinse the vine leaves, pat them dry on absorbent kitchen paper and lay them together in overlapping pairs. Brush the leaves with olive oil and divide the cheese mixture between them, piling it neatly on the middle of each pair. Wrap the edges of the leaves over to enclose the filling and secure with wooden cocktail sticks. Brush again with olive oil.
3 Place the parcels on the barbecue and cook for 2 to 3 minutes on each side, until charred. Remove the cocktail sticks and serve.

Vegetables

As you would expect, most of the recipes in this book are by their nature based on vegetables; however, I have grouped those in this chapter together as they have all been created for the main course. Of course, it is also easy to mix and match recipes as you wish, serving them in any combination you fancy for the main part of the meal.

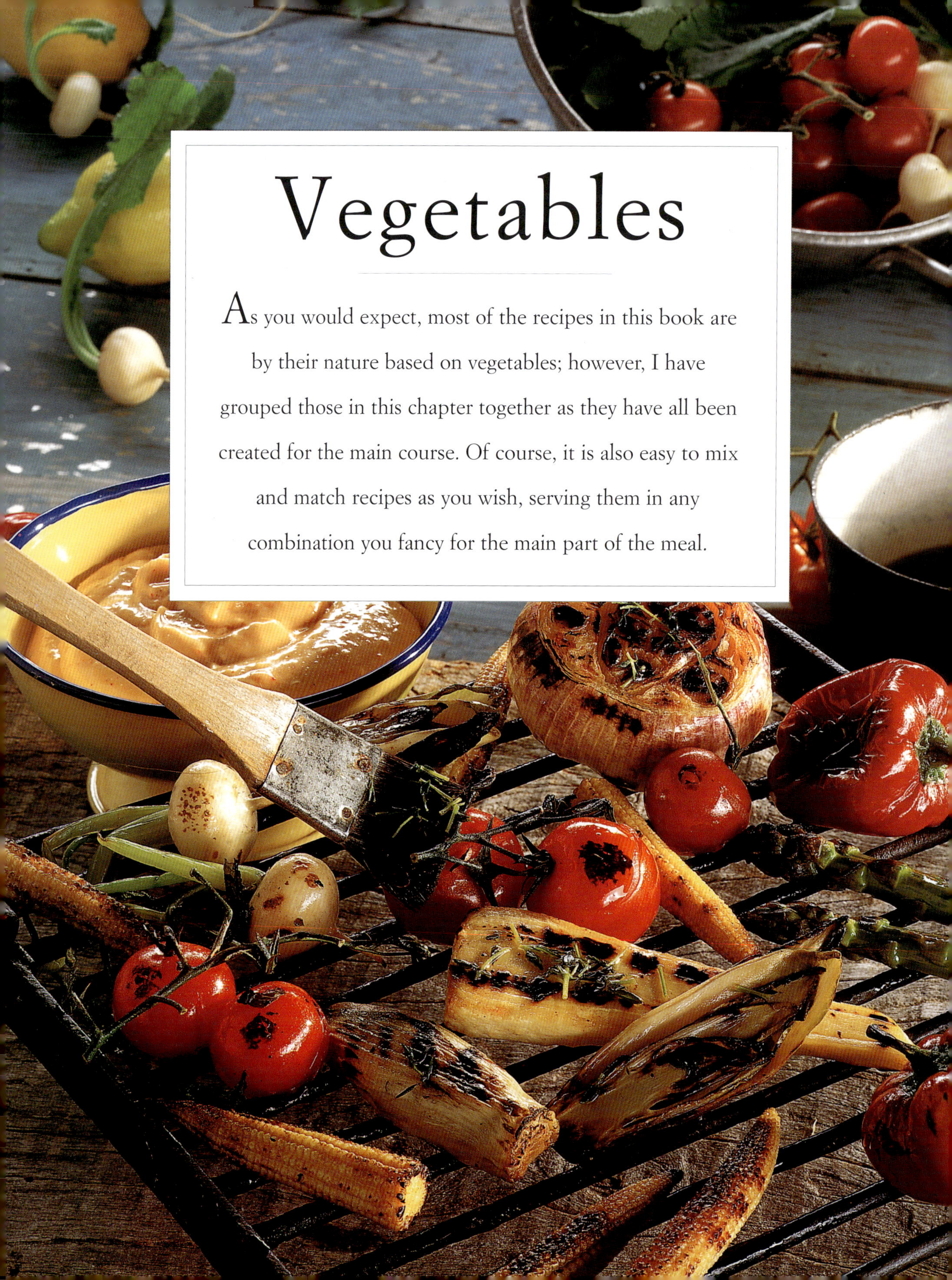

MIXED BABY VEGETABLE PLATTER WITH CHILLI MAYONNAISE

Time to prepare: 15 minutes
Time to cook: 12 to 18 minutes

Serves 4

4 tablespoons olive oil
2 tablespoons chopped fresh thyme
1.25 kg (2½ lb) mixed baby vegetables, for example courgettes, artichokes, asparagus tips, aubergines, baby plum tomatoes, carrots and parsnips
coarse sea salt

Chilli Mayonnaise
½ –1 medium red chilli
1 garlic clove
1 egg yolk
155 ml (5 fl oz) olive oil
juice of ½ lemon
1 teaspoon paprika
salt and freshly ground black pepper

Baby vegetables, char grilled and arranged on a large platter, make a wonderful centrepiece for a more formal barbecue. Serve with the chilli mayonnaise and mop up the juices with chunks of an Italian bread, such as focaccia. They can also be handed around as a starter, with bowls of extra virgin olive oil for dipping, or even tossed with freshly cooked linguine pasta and topped with a dusting of freshly grated Parmesan cheese, then served as a pasta course. If you choose the latter, leave out the chilli mayonnaise and just drizzle over some of your best extra virgin olive oil.

1 Place the olive oil in a small pan with the thyme and warm the mixture gently for 2 to 3 minutes. Set aside to infuse.
2 Prepare the chilli mayonnaise. Seed and finely chop the chilli, place it in a mortar and use a pestle to work it to a paste with the garlic. Place the egg yolk in a bowl and beat in the chilli paste. Slowly drizzle in the olive oil, whisking continuously, until the mixture thickens. Stir in the lemon juice, paprika and seasoning to taste. Transfer to a small serving bowl and set aside.
3 Trim and wash the vegetables according to type, cutting any larger vegetables in half lengthways.
4 Arrange the vegetables on the hot barbecue rack and brush with the infused oil. Cook for about 10 to 15 minutes, turning occasionally, until browned and tender.
5 Arrange the cooked vegetables on a platter, sprinkle with sea salt and serve at once, with the mayonnaise.

Brushing the mixed baby vegetables with thyme-infused oil

STUFFED FLAT MUSHROOMS WITH PINE NUTS AND OLIVES

Time to prepare: 15 minutes
Time to cook: 10 to 18 minutes

Serves 4

8 large open field mushrooms
60 g (2 oz) black olives, stoned
1 garlic clove
30 g (1 oz) pine nuts
30 g (1 oz) fresh white breadcrumbs
2 tablespoons freshly grated Parmesan cheese
2 tablespoons chopped fresh parsley
1 tablespoon chopped fresh marjoram
3 tablespoons olive oil
salt and freshly ground black pepper

Large flat mushrooms are easy to stuff and, when the filling is pressed in firmly, it is possible to turn the mushrooms without losing too much of the breadcrumb mixture. Try to pick mushrooms that have not opened out too much, so that the upturned edges will hold the stuffing in place.

1 Remove the stalks from the mushrooms, wipe the caps and set them aside. Place the stalks in a food processor with the olives, garlic and pine nuts, and process until finely chopped. Alternatively, finely chop these ingredients by hand.
2 Add the breadcrumbs, Parmesan, parsley, marjoram and 1 tablespoon of the oil. Mix well and season to taste.
3 Spoon the stuffing into the centres of the mushroom caps and press down firmly, smoothing the top of the stuffing.
4 Brush the top of each mushroom with some of the remaining oil and place on the hot barbecue, stuffing side down. Cook for 5 minutes, then turn the mushrooms very carefully and cook for a further 5 to 8 minutes, until cooked through. Serve immediately.

ROASTED BUTTERNUT SQUASH WITH ORANGE PECAN BUTTER

Time to prepare: 5 minutes
Time to cook: 22 to 25 minutes

Serves 2

1 butternut squash, about 750 g (1½ lb) in weight
40 g (1½ oz) softened butter
grated rind and juice of ½ orange
few drops Tabasco
2 tablespoons chopped fresh flat-leaf parsley
salt and freshly ground black pepper
15 g (½ oz) pecan halves

A variety of weird and wonderful squashes are now readily available in supermarkets and they are ideal for late summer barbecues. Squashes can taste rather bland, but when brushed with a flavoured butter and cooked over coals, their sweet flesh is transformed into a delicacy. I have used the golden, pear-shaped butternut squash here, but you can substitute other varieties, such acorn squash.

1 Place the whole squash in a large saucepan, pour over boiling water and bring back to the boil. Simmer for 10 minutes.
2 Place the butter in a small bowl and beat in the orange rind and juice, Tabasco, parsley and plenty of seasoning. Finely chop the pecans and beat into the butter.
3 Drain the squash and cut it in half. Scoop out and discard the seeds. Place the squash halves on the barbecue, cut sides down, and cook for 5 minutes. Turn the squash over and spread the flavoured butter over the cut sides. Cook for a further 7 to 10 minutes, until the flesh is tender. Serve, piping hot, with a mixture of wild and long-grain rice, and a salad of bitter leaves.

PATATAS BRAVAS

Time to prepare: 10 minutes
Time to cook: 38 to 39 minutes

Serves 4

12 large new potatoes
1 small onion
1 garlic clove
1 small dried red chilli
4 tablespoons olive oil
250 g (8 oz) passata
6 tablespoons dry white wine
2 tablespoons chopped flat-leaf parsley
salt and freshly ground black pepper
1/2 teaspoon paprika
flat-leaf parsley, to garnish

This is a version of one of my favourite tapa, encountered in Seville during my first visit to mainland Spain, when I fell in love with the country, its food and people. One particular small bar, situated across the Qualdquiver river in the gypsy quarter, became a favourite haunt; I can think of nothing more relaxing than sitting on a bar stool after a morning's sight seeing, picking at these piquant little potatoes while sipping glasses of chilled fino sherry. Although the total cooking time is slightly over 30 minutes, the potatoes and sauce simmer simultaneously, so the dish can be cooked within 30 minutes.

1 Scrub the potatoes, place them in a saucepan of cold water and bring to the boil. Cook for 10 minutes, until the potatoes are almost tender.
2 While the potatoes are cooking, prepare the tomato sauce. Finely chop the onion and garlic, and crush the chilli.
3 Heat 1 tablespoon of the oil in a small saucepan and add the onion and garlic. Cook for 3 minutes, until softened, then add the chilli, passata, wine, parsley and seasoning. Simmer for 20 minutes until thickened, stirring occasionally.
4 Drain the potatoes and refresh them under cold water. Pat dry and cut each one into four wedges. Brush the potato wedges with the remaining olive oil and sprinkle with the paprika. Arrange on the barbecue and cook for 5 to 6 minutes, turning frequently, until golden all over.
5 Transfer the potatoes to a serving dish, pour the tomato sauce over them. Garnish with flat-leaf parsley and serve.

CHAR-GRILLED FENNEL, COURGETTE AND ONION RICE

Time to prepare: 10 minutes
Time to cook: 15 to 20 minutes

Serves 4

185 g (6 oz) basmati rice
2 fennel bulbs
2 medium courgettes
1 large Spanish onion
5 tablespoons extra virgin olive oil
1 garlic clove
grated rind and juice of ½ lemon
2 tablespoons chopped chives
2 tablespoons chopped fresh flat-leaf parsley
1 teaspoon ground cumin
salt and freshly ground black pepper
60 g (2 oz) baby plum tomatoes

I like to use brown rice for this recipe but that takes slightly longer than 30 minutes to cook, so I have used basmati rice instead which adds its own fragrant character to the finished dish. If you do have the time, Italian easy-cook brown rice has a delicious nutty flavour that I particularly recommend for this warm rice salad.

1 Bring a large saucepan of water to the boil, add the rice and cook for 10 minutes.
2 Meanwhile, trim the fennel and cut it into quarters, slice the courgettes lengthways, and thickly slice the onion. Brush the vegetables with 2 tablespoons of the oil and place on the hot barbecue. Cook for 10 to 15 minutes, turning regularly, until charred. If necessary, cook the fennel for a little longer.
3 Drain the rice and place it in a serving bowl. Crush the garlic and whisk it with the remaining oil, the lemon rind and juice, chives, parsley, cumin and seasoning. Pour this dressing over the rice and mix well.
4 Cut the vegetables into large pieces and halve the tomatoes, then stir them into the rice. Check the seasoning before serving warm.

POTATO WEDGES WITH ROSEMARY AND GARLIC

Time to prepare: 10 minutes
Time to marinate: 30 minutes
Time to cook: 15 to 20 minutes

Serves 4

750 g (1½ lb) even-sized waxy potatoes
4 tablespoons extra virgin olive oil
2 large fresh rosemary sprigs
2 garlic cloves
salt and freshly ground black pepper

This is an old favourite that I have transferred from the oven to the barbecue with great success. Hand the potato wedges around with Salsa Verde, see page 79, as a dip before the main course, or serve them with char-grilled peppers and aubergines drizzled with a little balsamic vinegar.

1 Scrub the potatoes and cut them in half lengthways. Cut each potato half into three thin wedges. Place in a shallow dish and pour over the olive oil.
2 Chop the rosemary and crush the garlic, then add both to the potatoes with plenty of seasoning. Toss well to thoroughly coat in the oil. Set the potatoes aside to marinate for about 30 minutes.
3 Remove the potato wedges from the marinade and arrange in a hinged grill basket to make turning easier. Place on the barbecue and cook for 15 to 20 minutes, turning regularly, until the potatoes are golden and tender. Brush the potatoes with any remaining oil as they cook. Serve immediately.

Cook's Tip
When they are available, use small new potatoes for this dish. I usually thread them onto metal skewers to make them easier to turn on the barbecue.

PATTY PAN SQUASHES WITH LEMON MUSTARD DRESSING

Time to prepare: 5 minutes
Time to cook: 10 to 15 minutes

Serves 4

750 g (1½ lb) patty pan squashes (buy a mixture of colours, if possible)
grated rind and juice of 1 lemon
1 tablespoon grainy mustard
1 tablespoon light muscovado sugar
1 tablespoon chopped fresh oregano
5 tablespoons olive oil
salt and freshly ground black pepper
30 g (1 oz) black olives, stoned
250 g (8 oz) mixed salad leaves

Like their close relation, the courgette, these little squashes get very waterlogged when overcooked. Barbecuing them with a well-flavoured sauce or, as in this case, an oil-based dressing, avoids this problem. The dressing also acts as a basting sauce for the squashes while they are cooking. Patty pan squashes are also ideal for kebabs and stir fries. The Tomato Chilli Relish, see page 84, would go well with this dish

1 Wash the squashes and thread them onto skewers. Place the lemon rind and juice, mustard, sugar, oregano and oil in a screw-topped jar with plenty of seasoning and shake well until emulsified.
2 Brush some of the dressing over the squashes and cook them on the hot barbecue for 10 to 15 minutes, turning regularly, until tender and browned.
3 Roughly chop the olives, place in a bowl with the salad leaves and toss with a little of the dressing. Arrange in the base of a serving dish.
4 Remove the squashes from the skewers and toss them with the rest of the dressing until coated. Arrange on the bed of leaves and serve at once.

Cook's Tip
When picking patty pans, look for vegetables of a good, even colour, with smooth taut skins that are free of surface blemishes.

SWEET POTATOES WITH QUICK CURRY PASTE

Time to prepare: 10 minutes
Time to cook: 15 to 20 minutes

Serves 4

750 g (1½ lb) sweet potatoes

Quick Curry Paste

1 tablespoon cumin seeds
1 teaspoon coriander seeds
½ teaspoon black peppercorns
1 tablespoon paprika
½ teaspoon cayenne
1 tablespoon ground turmeric
1 teaspoon salt
2 garlic cloves
juice of 1 lemon
4 tablespoons natural yogurt

I can think of no better way of cooking a sweet potato than over hot coals. The whole process is simplicity itself and it intensifies the nutty flavour of the vegetables while cooking the outside to a crisp shell. For this recipe, simply slice the sweet potatoes and brush with the wonderful spicy paste. I tend to make up a jar of paste and keep it in the fridge as it can be used in all kinds of dishes.

1 First prepare the curry paste. Heat a small frying pan for a couple of minutes over a high heat then add the cumin, coriander and peppercorns and cook for 30 seconds.
2 Transfer the roasted spices to a mortar and grind to a fine powder with a pestle. Work in the paprika, cayenne, turmeric and salt. Then add the garlic cloves and a little lemon juice and pound the mixture to a paste. Stir this paste into the yogurt with the rest of the lemon juice.
3 Scrub the sweet potatoes and cut them into 1 cm (½ inch) thick slices. Pour the curry paste over the slices and mix well until coated.
4 Arrange the sweet potato slices on the hot barbecue and cook for 15 to 20 minutes, turning regularly, until tender. Serve at once, with mango chutney.

Cook's Tip

This method also works well with other root vegetables, such as potatoes, swede and turnips.

TORTILLAS WITH SWEETCORN AND AVOCADO

Time to prepare: 15 to 20 minutes
Time to marinate: 1 hour
Time to cook: 12 to 15 minutes

Serves 4

4 corn cobs
2 tablespoons tomato ketchup
2 tablespoons vegetarian Worcestershire sauce
few drops Tabasco
2 tablespoons light muscovado sugar
juice of ½ lime
1 red onion
½ green pepper
2 large avocados
4 ready-made flour tortillas
2 tablespoons soured cream or fromage frais
Hot Tomato and Pepper Salsa, see page 94, to serve

Look out for soft flour tortillas in your local supermarket – they warm up quickly on the barbecue and can be used as an easy way to serve all kinds of roasted vegetables. Simply wrap the chosen filling in a tortilla and eat it. I suggest marinating the corn for 1 hour, but this can be reduced or omitted when time does not allow for marinating.

1 Remove the husks from the corn. Bring a large saucepan of water to the boil and cook the corn for 5 minutes. Drain the corn cobs thoroughly.
2 Mix the tomato ketchup, Worcestershire sauce, Tabasco, sugar and lime juice. Pour over the corn cobs and leave to marinate for 1 hour, turning occasionally.
3 Slice the onion and seed and slice the pepper. Cook the corn cobs on the hot barbecue for 5 to 8 minutes, turning occasionally, until charred and golden. Meanwhile, halve, stone and skin the avocado. Slice the flesh and toss it in any remaining marinade.
4 Holding the hot cobs firmly with a cloth, run a knife down the length of each to remove the corn. Place in a bowl and mix with the onion, peppers and avocado.
5 Warm the tortillas for about 2 minutes on the barbecue. Divide the corn and avocado mixture between the warm tortillas. Spoon a dollop of soured cream or fromage frais on top and roll up the tortillas loosely. Serve at once, with Hot Tomato and Pepper Salsa.

Cook's Tip

Scraping the corn from the cobs is very fiddly and you may just want to serve the whole cobs to your guests, leaving them to scrape off the kernels, if they wish. I recommend that you have a go, if you can be bothered, as the charred corn goes so well with the cool creamy avocado, all wrapped in an easy-to-eat parcel.

Salads

Any barbecued vegetable, tossed with a little good olive oil and a dash of balsamic vinegar, makes an enticing salad; serve the salad at room temperature with sour-dough bread or olive-oil bread and you have the base for a satisfying meal. Most of these salads will improve in flavour if you leave them to stand for at least 24 hours.

COUSCOUS SALAD WITH BARBECUED VEGETABLES

Time to prepare: 15 minutes
Time to cook: about 15 minutes

Serves 4

250 g (8 oz) couscous
375 ml (12 fl oz) cold water or vegetable stock
60 g (2 oz) butter
1/2 teaspoon ground cinnamon
1/2 teaspoon ground ginger
1/2 teaspoon paprika
8 shallots
1 aubergine
2 medium courgettes
1 yellow pepper
1 garlic clove, crushed
1 tablespoon balsamic vinegar
3 tablespoons chopped fresh coriander
2 tablespoons chopped fresh parsley
salt and freshly ground black pepper

Brushing a spicy butter over the cooking vegetables gives this salad a wonderful, aromatic sweet-and-sour flavour. Serve it piled into warm pitta bread, with Barbecued Halloumi Strips, see page 37, for a complete meal that can be eaten without knives and forks – great finger food for a beach barbecue.

1 Place the couscous in a bowl and pour in the water or stock. Leave to stand for 15 minutes, stirring occasionally to break up any lumps.
2 Gently heat the butter and spices together in a small saucepan until melted. Set aside.
3 Place the shallots in a bowl and pour over boiling water; leave to stand for 5 minutes, then drain and peel them. Thread the shallots onto skewers.
4 Thinly slice the aubergine and courgettes lengthways. Halve and seed the pepper. Place all the vegetables on the barbecue and brush with the spice butter. Cook the vegetables for 10 minutes, brushing with butter and turning, until they are tender and just blackened.
5 Remove the shallots from the skewer and cut in half. Cut the aubergine, pepper and courgettes into thin strips.
6 Add the vegetables to the couscous with any remaining butter. Stir in the garlic, balsamic vinegar, coriander, parsley and seasoning. Serve warm or cold.

Cook's Tip
For a more substantial salad add 125 g (4 oz) cubed feta or a firm goats' cheese to the couscous along with the barbecued vegetables.

Pouring the remaining spiced butter over the Couscous Salad with Barbecued Vegetables

WARM PUY LENTIL SALAD WITH CHAR-GRILLED AUBERGINE AND FENNEL

Time to prepare: 15 minutes
Time to cook: 35 to 45 minutes

Serves 4

185 g (6 oz) Puy lentils
2 fennel bulbs
1 large aubergine
1 red onion
30 g (1 oz) pine nuts
1 garlic clove
2 tablespoons chopped fresh mint
1 teaspoon ground cumin
4 tablespoons extra virgin olive oil
1 tablespoon balsamic vinegar
salt and freshly ground black pepper

Puy lentils are the small slate-green pulses that come from the area around the French town of Le Puy. Regarded by chefs as the king of the lentils for their fine texture, smoky flavour and ability to hold their shape, Puy lentils make a subtle partner for the rich flavours of barbecued aubergine and fennel. Start to cook the lentils 15 minutes before putting the vegetables on the barbecue and the dish will be ready within 30 minutes in spite of the slightly longer total cooking time.

1 Put the lentils in a saucepan with enough cold water to cover them generously. Bring to the boil and simmer for 25 to 30 minutes, until just tender. Drain and place in a salad bowl.
2 While the lentils are cooking, trim the fennel and cut the bulbs into thin slices from tip to root. Cut the stalk from the aubergine and cut it into 1 cm (1/2 inch) thick slices. Slice the onion. Toast the pine nuts in a dry frying pan until golden. Crush the garlic.
3 Cook the fennel and aubergine on the barbecue for 10 to 15 minutes, until charred and tender. Cut the grilled vegetables into thin strips and add them to the warm lentils with the onion, pine nuts and garlic.
4 Mix in the mint, cumin, olive oil, balsamic vinegar and seasoning. Serve at once.

Cook's Tip
Do not try to use red or brown lentils in this dish as they tend to go mushy and the results will be very disappointing.

GRILLED BREAD SALAD WITH PARMESAN AND MUSHROOMS

Time to prepare: 10 minutes
Time to cook: 4 to 5 minutes

Serves 4

4 thick slices olive-oil bread, such as ciabatta
1 garlic clove
375 g (12 oz) small chestnut mushrooms
4 tablespoons chopped fresh flat-leaf parsley
4 tablespoons extra virgin olive oil
1 tablespoon balsamic vinegar
salt and freshly ground black pepper
60 g (2 oz) Parmesan

Char grilling the bread creates a hearty, smoked flavour that marries well with the mild flavour of the mushrooms in this salad.

1 Toast the bread on the barbecue for 1 to 2 minutes on each side. Cut the garlic clove in half and rub it over one side of each slice of bread. Cut the bread into small cubes and place in a salad bowl.
2 Wipe the mushrooms and cut them into quarters. Add to the bowl with the parsley. Whisk the olive oil and vinegar with plenty of seasoning and then drizzle this dressing over the salad.
3 Shave the Parmesan into thin flakes with a vegetable peeler and scatter them over the salad. Lightly toss the ingredients together and serve.

Cook's Tip
This salad can be made with a mixture of mushrooms, such as shiitake and oyster mushrooms.

ROCKET WITH CHAR-GRILLED RED PEPPER AND CARAMELISED SHALLOTS

Time to prepare: 10 minutes
Time to cook: about 15 minutes

Serves 4

8 shallots
2 red peppers
30 g (1 oz) pine nuts
60 g (2 oz) feta cheese
90 g (3 oz) rocket leaves
1 tablespoon red wine vinegar
1 tablespoon light muscovado sugar
1 tablespoon olive oil plus 2 tablespoons extra virgin olive oil
salt and freshly ground black pepper

The colours of this salad are vibrant and satisfying. Rocket is now available in many supermarkets, in sealed bags or growing in little pots, but it is also incredibly easy to grow, either in a pot or in a small flower bed. It bolts, or goes to seed, rapidly, like spinach, so you have to keep picking it or pinching off the top buds and leaves to keep the plant compact and leafy.

1 Place the shallots in a bowl, pour over boiling water and leave for 2 minutes. Then drain and peel the shallots, and thread them onto skewers. Halve and seed the peppers.
2 Toast the pine nuts in a small frying pan over high heat until golden. Cube the feta. Mix the pine nuts, feta and rocket in a salad bowl.
3 Mix the vinegar, sugar and 1 tablespoon olive oil, and brush this over the shallots and peppers. Cook the shallots and peppers on the barbecue for 10 to 15 minutes, brushing with the marinade and turning, until the shallots are golden and the skins of the peppers are blackened.
4 Place the peppers in a plastic bag and set them aside for 5 minutes to loosen their skins. Remove the shallots from the skewers. Peel the peppers and cut them into strips.
5 Arrange the vegetables on the salad. Drizzle the extra virgin olive oil over and season to taste. Toss gently together and serve with bought bread or with Potato and Onion Bread, see page 76.

ROASTED RATATOUILLE SALAD

Time to prepare: 15 minutes
Time to cook: 15 minutes

Serves 4

1 large aubergine
2 medium courgettes
2 red peppers
3 ripe tomatoes
1 large mild onion
4 tablespoons extra virgin olive oil
2 garlic cloves
2 tablespoons red wine vinegar
1 tablespoon capers
salt and freshly ground black pepper
chopped fresh basil, to garnish

Ratatouille can be a disappointing mess if all the vegetables are overcooked and the flavours and textures blend together into a total mush. Roasting the vegetables keeps their flavours distinct and here the capers and red wine vinegar add a welcome piquancy. This can only be cooked within 30 minutes if you have a very large barbecue, otherwise you will have to cook the vegetables in batches.

1 Trim the aubergine and courgettes and slice them into thin rounds. Halve and seed the peppers. Halve the tomatoes and onion. Reserve half the olive oil and use the rest to brush over the vegetables.
2 Cook the vegetables on the barbecue for 15 minutes, until tender and blackened. Place the peppers and tomatoes in a plastic bag and leave for 5 minutes to loosen their skins, then peel them and cut into strips. Slice the onion halves so that the slices separate into strips.
3 Place all the vegetables in a salad bowl. Crush the garlic and add it to the salad with the vinegar, capers, seasoning and the remaining olive oil. Scatter with chopped basil and serve at once.

Cook's Tip
This salad makes a tempting topping for a pizza. Spread it over a couple of prepared pizza bases and scatter some grated mozzarella over the top. Cook for 10 to 12 minutes in a hot oven.

ROASTED PLUM TOMATOES WITH CHICK PEAS

Time to prepare: 15 minutes
Time to cook: 25 to 30 minutes

Serves 4

1 whole head of garlic
500 g (1 lb) plum tomatoes
400 g (14 oz) can chick peas, drained
1 medium red onion
6 tablespoons chopped fresh basil
4 tablespoons extra virgin olive oil
salt and freshly ground black pepper

Visiting Sicily had a profound effect on the way I cook. The intensity of the flavours, colours and scents of the island's cuisine have evolved from a diverse cultural heritage – the Moors, Normans, Spanish and French are just some of the peoples who have conquered the island and moved on, leaving behind their influence, evident in today's culinary style. Couscous, saffron and spices; fine pastries with almonds and rose water; and a wide selection of superb fruit and vegetables all feature in Sicilian cooking. This salad is a version of one I ate during my stay; it was served with barbecued squid and seafood couscous, but it was so good that I could have eaten it on its own, with the local bread to mop up the juices.

1 Wrap the head of garlic in foil and cook it among the barbecue coals for 15 to 20 minutes, until tender.
2 Meanwhile, cook the tomatoes on the barbecue for about 10 minutes, until the skins blacken. Place in a plastic bag and leave for 5 minutes.
3 While the tomatoes are cooking, place the chick peas in a salad bowl. Slice the onion and add to the chick peas with the basil, oil and seasoning.
4 Peel the tomatoes and chop the flesh. Add to the salad and mix well. Squeeze the flesh from the garlic cloves into the salad, mix and leave to stand until ready to serve.

FENNEL AND RED ONION SALAD

Time to prepare: 15 minutes
Time to cook: 10 minutes

Serves 4

2 large fennel bulbs
2 large oranges
1 medium red onion
3 tablespoons extra virgin olive oil
1 tablespoon balsamic vinegar
1 tablespoon chopped fresh mint
salt and freshly ground black pepper
mint, to garnish

Fennel is an ideal vegetable for barbecue cooking. Cut into thin slices, its strong aniseed flavour mellows as it char grills. Juicy oranges and sweet red onions are excellent partners for the fennel. For those who eat fish, this salad goes well with mackerel or tuna.

1 Trim the root ends off the fennel and cut the bulbs lengthways into 5 mm (1/4 inch) thick slices. Cook on the barbecue for 5 minutes on each side, turning once, until tender and blackened.
2 While the fennel is cooking, carefully peel the oranges with a serrated knife, taking care to remove all the white pith. Cut the flesh into thin slices. Thinly slice the onion.
3 Whisk together the olive oil, balsamic vinegar, mint and seasoning. Layer the fennel in a serving dish with the oranges and onion, drizzling each layer with dressing. Serve at room temperature, garnished with mint.

Cook's Tip
If preparing the fennel in advance, place it in a bowl of water with the juice of 1/2 lemon to prevent it from discolouring. As the slices can fall apart, either hold them in place with wooden cocktail sticks or hold them in a grilling rack before placing on the barbecue.

SALAD OF STUFFED PEPPERS WITH TOMATO AND FENNEL

Time to prepare: 10 minutes
Time to cook: 10 to 15 minutes

Serve 4

2 yellow peppers
2 red peppers
1 medium fennel bulb
8 baby plum or cherry tomatoes
3 tablespoons extra virgin olive oil
salt and freshly ground black pepper
2 tablespoons pesto

Stuffed peppers make a colourful dish that can be served either hot or at room temperature, so they are ideal if your barbecue is getting a little overcrowded as they can be cooked, then set aside. Securing the stuffing in place with cocktail sticks ensures that most of it remains in place as the peppers cook, but remember to soak the sticks in water first to prevent them from burning.

1 Wash the peppers and cut in half lengthways. Remove the seeds. Cut the fennel into thin slices.
2 Place two tomatoes into each pepper half and arrange slices of fennel on top. Secure the fennel in place with a couple of wooden cocktail sticks placed diagonally across the top of each pepper half. Brush the peppers all over with olive oil and season with salt and pepper.
3 Place the peppers cut sides down on the barbecue and cook for 5 minutes. Turn and continue cooking for about 10 minutes, until the peppers are just charred and tender.
4 Transfer the peppers to a serving dish. Remove the cocktail sticks and leave until just cool. When the peppers are still slightly warm, drizzle each one with a little pesto and serve with plenty of bread to mop up any juices.

Cook's Tip
Strict vegetarians can use pesto that has been made with a vegetarian Parmesan, or substitute a good-quality bought olive paste instead.

NEW POTATOES AND CAULIFLOWER WITH MINT

Time to prepare: 10 minutes
Time to cook: 15 to 20 minutes

Serves 4

500 g (1 lb) small new potatoes (Jersey Royals if possible)
1 small cauliflower
4 tablespoons olive oil
2 tablespoons chopped fresh mint
juice of 1/2 lemon
salt and freshly ground black pepper
1 teaspoon curry paste

A potato salad may seem a little dull, but made with small Jersey Royal potatoes and enlivened with barbecued cauliflower dressed with mint, this becomes a meal fit for a king. The hint of spice from good-quality curry paste unites the flavours and textures. Serve in warm naan bread, with a fresh tomato and onion chutney and yogurt raita.

1 Cook the potatoes in boiling water for 10 to 15 minutes until tender. Cut the cauliflower into florets and cook in boiling water for 5 minutes until almost tender. Drain both vegetables and place the potatoes in a salad bowl.
2 Whisk together 3 tablespoons of the oil, the mint, lemon juice and seasoning. Pour this dressing over the potatoes and toss to coat them.
3 Mix the remaining oil with the curry paste. Arrange the cauliflower florets on the barbecue and brush them with the flavoured oil. Cook for 5 minutes, turning and brushing with more oil. Toss the cauliflower with the potatoes and serve at room temperature.

Cook's Tip
This recipe works without barbecuing the cauliflower, but do try it this way as the smoky taste imparted by the coals is delicious with the vegetables.

Breads
and accompaniments

The breads in this chapter are all incredibly simple to make and cook, yet they will impress your guests who will feel that you have made a special effort for them — just the effect I try to achieve with all my cooking!

POLENTA AND PARMESAN WEDGES

Time to prepare: 15 minutes
Time to cook: 11 to 20 minutes

Serves 4

30 g (1 oz) pine nuts
8 black olives, stoned
750 ml (24 fl oz) water
salt and freshly ground black pepper
185 g (6 oz) instant polenta
15 g ($^1/_2$ oz) butter
30 g (1 oz) freshly grated Parmesan cheese
3 tablespoons chopped fresh basil
2 tablespoons chopped fresh flat-leaf parsley
2 tablespoons olive oil

On its own, polenta can be uninspiring to say the very least! Flavoured with Parmesan and fresh herbs, and studded with pine nuts and black olives, it becomes something else entirely, especially when cooked over coals to give it a delicious, earthy flavour and crisp exterior. Serve these wedges with Tomato Chilli Relish, see page 84, or one of the salsas.

1 Place the pine nuts in a small frying pan and cook over medium heat for 1 to 2 minutes, until golden. Chop the olives. Set both aside.
2 Bring the water to the boil in a large saucepan. Add $^1/_2$ teaspoon salt and pour in the polenta, stirring continuously. Simmer over low heat for 5 to 8 minutes, stirring all the time, until the polenta is thick and coming away from the sides of the pan.
3 Stir the butter, Parmesan, basil, parsley, olives and toasted pine nuts into the polenta and season to taste. Spoon the mixture into a buttered roasting tin and spread it out to a thickness of 2.5 cm (1 inch). Leave to cool for 10 minutes, then cut into triangles or wedges.
4 Brush the polenta with olive oil and cook it on the hot barbecue for 5 to 10 minutes, turning once or twice, until golden and heated through. Serve with a salsa.

Lifting Polenta and Parmesan Wedges off the grid, ready to serve with Tomato Chilli Relish, *see page 84*

SAGE SHEET BREAD

Time to prepare: 10 minutes
Time to cook: 15 to 20 minutes

Serves 4

155 g (5 oz) plain flour
90 g (3 oz) fine semolina
1 teaspoon salt
1 tablespoon chopped fresh sage
200 ml ($6^1/_2$ fl oz) warm water
2 tablespoons extra virgin olive oil
sea salt

I was introduced to sheet bread by cookery writer Rosemary Stark. It originates from Sardinia, where it is also called *Carta da Musica* because it looks like old sheet music. It is simple to prepare and cooks perfectly on the barbecue, so I make up the dough in advance and cook a few sheets to serve with dips and salsas while the main course is grilling.

1 Sift the flour into a large mixing bowl and stir in the semolina, salt and sage. Pour in the water and mix to a soft dough. Turn the dough onto a floured surface and knead it for 1 minute, adding more flour if the dough gets too sticky.
2 Divide the dough into eight and cover with a damp cloth to prevent it from drying out while you work.
3 Roll out a portion of dough on a floured surface, into a very thin circle measuring about 23 cm (9 inch) in diameter.
4 To cook the sheet bread, lay it on the hot barbecue and cook for 1 to 2 minutes on each side, until the surface puffs and bubbles. Brush with olive oil and sprinkle with sea salt while still warm. Continue rolling out and cooking the rest of the dough. Serve the bread warm.

GARLIC HERB PIZZA BREAD

Time to prepare: 10 minutes
Time to cook: 8 to 10 minutes

Serves 4

60 g (2 oz) butter
2 garlic cloves
2 tablespoons chopped fresh flat-leaf parsley
salt and freshly ground black pepper
250 g (8 oz) packet pizza-base mix
1 tablespoon extra virgin olive oil

There are plenty of good pizza-base mixes available now and they are ideal for this barbecue version of garlic bread. It tastes wonderful, so be sure to make enough to satisfy everyone, especially children. My children tend to get to this first and it disappears rapidly, before adults have time to eat their fair share.

1 Place the butter in a small bowl and beat it until softened. Finely chop the garlic and beat it into the butter with the parsley and plenty of seasoning.
2 Make up the pizza-base mix as instructed on the packet, adding the olive oil with the water. Divide the dough in half and roll out two circles, each about 5 mm (1/2 inch) thick. Do not roll the dough too thin as it will sink through the grill. Use a sharp knife to slash the top of each dough circle about five times.
3 Lay the dough circles on the well-oiled barbecue grill. Cook for 4 to 5 minutes on each side, until golden and crisp. Spread the bread with the garlic butter, cut the circles into strips and serve immediately.

GOATS' CHEESE AND DILL SCONES

Time to prepare: 10 minutes
Time to cook: 12 to 15 minutes

Makes 10 to 12 scones

250 g (8 oz) self-raising flour
1 teaspoon salt
large pinch cayenne pepper
½ teaspoon baking powder
30 g (1 oz) butter
1 tablespoon chopped fresh dill
60 g (2 oz) goats' cheese
155 ml (5 fl oz) natural yogurt
beaten egg to glaze

I serve these scones topped with dollops of mascarpone and strips of barbecued vegetables tossed with a little balsamic vinegar. Use a firm, crumbly goats' cheese rather than any of the soft fresh types.

1 Preheat the oven to 200°C, 400°F, Gas 6. Place a baking sheet in the oven to heat.
2 Sift the flour, salt, cayenne and baking powder into a large mixing bowl. Cut the butter into dice and rub it into the dry ingredients until the mixture resembles fine breadcrumbs. Stir in the dill. Crumble the goats' cheese and stir it in. Add the yogurt and mix to a soft dough.
3 Turn the dough onto a lightly floured work surface and knead lightly. Roll out to 2.5 cm (1 inch) thick and use a plain cutter to cut out rounds. Roll out the trimmings and cut more scones. Place on the baking sheet and brush the tops with beaten egg.
4 Bake for 12 to 15 minutes, until well risen and golden. Transfer to a wire rack to cool.

Cook's Tip
These scones can be made with alternative flavourings; for example, try replacing the dill and goats' cheese with basil and feta cheese or sun-dried tomato and mozzarella cheese.

POTATO AND ONION BREAD

Time to prepare: 15 minutes
Time to cook: 25 minutes

Makes 1 loaf

1 medium mild onion
1 tablespoon olive oil
250 g (8 oz) strong plain flour
2 teaspoons baking powder
1 teaspoon salt
1/4 teaspoon paprika
185 g (6 oz) cooked floury potatoes
1 egg
155 ml (5 fl oz) semi-skimmed milk
30 g (1 oz) melted butter
1 teaspoon caraway seeds

This hearty bread makes a wonderful accompaniment for barbecued vegetables, particularly sweet varieties, such as peppers and parsnips. The bread is best when fresh, so make up the dough as soon as the barbecue is lit. It will cook in half an hour and still be warm and delicious when the rest of your meal is ready.

1 Preheat the oven to 230°C, 450°F, Gas 8. Finely chop the onion. Heat the oil in a medium frying pan and cook the onion for 5 minutes, until softened.
2 Meanwhile, sift the flour, baking powder, salt and paprika into a large mixing bowl. Rub the potatoes into the dry ingredients and stir in the onion.
3 Beat the egg with the milk, stir in the melted butter and add to the ingredients in the bowl. Mix to a soft dough, which should be quite wet.
4 Turn the dough out onto a well-floured work top and shape it into a 20 cm (8 inch) round. Place on a baking sheet and sprinkle with the caraway seeds. Cut into eight wedges and cook for about 25 minutes, until browned. When the bread is cooked, it will sound hollow when tapped on the base. Serve warm.

Cook's Tip
To make wonderful bruschetta, bake the bread in advance without cutting it into wedges before you cook it. Leave to cool then cut into thick slices. Cook these on the hot barbecue for 1 minute on each side, until toasted. Moisten with olive oil and serve with the chosen topping, for example the topping on the Roasted Tomato Crostini, see page 12, or on Charred Spring Onion Bruschetta, see page 13, are both suitable.

CINNAMON AND HONEY PANCAKES

Time to prepare: 10 minutes
Time to cook: 15 minutes

Serves 4

250 g (8 oz) self-raising flour
½ teaspoon ground cinnamon
½ teaspoon cream of tartar
¼ teaspoon bicarbonate of soda
¼ teaspoon salt
1 egg
315 ml (10 fl oz) semi-skimmed milk
2 tablespoons honey
a little sunflower oil for frying
30 g (1 oz) butter

My daughters are avid pancake fans and it is thanks to them that I discovered how good Scotch pancakes, or drop scones as they are also known, taste when heated through on the barbecue. We reheated the remains of a batch over the fading coals of a summer barbecue and served them with barbecued strawberries drizzled with honey and they were quite wonderful – well worth trying! Drop scones freeze well, so make and freeze a batch if you want to save time. I make the savoury version to serve with char-grilled vegetables and Salsa Verde, see page 79.

1 Sift the flour, cinnamon, cream of tartar, bicarbonate of soda and salt into a large mixing bowl. Beat the egg with the milk and honey. Make a well in the centre of the dry ingredients, add the liquid and gradually beat the dry ingredients into the liquid to make a smooth batter.
2 Heat a little oil in a large frying pan and spoon a large tablespoon of the mixture onto the hot surface. Cook for about 2 minutes, until just set and pale golden, then turn and cook the other side. Transfer to a plate. Cook the rest of the mixture in batches.
3 To finish the pancakes, melt the butter in a small pan and brush it over the pancakes. Place them on the hot barbecue and heat through on both sides. Serve with barbecued fruit.

Savoury Pancakes
For savoury pancakes, omit the cinnamon and honey from the recipe and add chopped fresh herbs to the batter: snipped chives, chopped thyme and/or tarragon are suitable.

Sauces and relishes

Well-flavoured sauces are essential for barbecue cooking: they provide welcome moisture, add zest to bland foods or a dash of colour and texture where needed. Serve them to enhance, but not overpower, the main ingredients.

ROMESCO SAUCE

Time to prepare: 10 minutes
Time to cook: 8 to 15 minutes

Serves 4

1 medium onion
3 garlic cloves
2 red peppers
4 ripe tomatoes
60 g (2 oz) blanched almonds or hazelnuts
1 large red chilli
2 tablespoons red wine vinegar
125 ml (4 fl oz) extra virgin olive oil
salt and freshly ground black pepper

This stunning yet simple sauce based on roasted vegetables and nuts, comes from the Tarragona region in Spain where it is traditionally served with barbecued shellfish. Grill the peppers and tomatoes over the barbecue to make a wonderfully smoky sauce that can be used as a dip or relish, or it can even be poured over pasta.

1 Halve the onion horizontally. Thread the garlic onto a wooden skewer.
2 Place the onion, garlic, whole peppers and tomatoes on the hot barbecue and cook for 8 to 15 minutes, turning regularly, until the skins are charred. The garlic and tomatoes will cook faster and should be removed from the barbecue first.
3 Place the peppers and tomatoes in a bowl, cover and set aside for 5 minutes until their skins have loosened.
4 Peel and seed the peppers and tomatoes, catching any juice. Roughly chop the flesh and place in a blender or food processor with any juice.
5 Dry fry the almonds or hazelnuts in a small frying pan for about 2 minutes, until golden. Add to the tomato mixture and process to a paste. Chop and seed the chilli, and add it to the sauce with the garlic and onion.
6 With the motor running, slowly pour in the vinegar and olive oil, then continue to process until the sauce is smooth. Season to taste and leave at room temperature until you are ready to serve the sauce.

Cook's Tip
Store any extra sauce in a covered container in the fridge for up to a week. Bring the sauce to room temperature before it is served to allow the flavours to develop.

Dry frying almonds ready for Romesco Sauce

HOT TOMATO AND PEPPER SALSA

Time to prepare: 10 minutes

Serves 4

250 g (8 oz) ripe plum tomatoes
1–2 green chillies
1 yellow pepper
1 small red onion
2 tablespoons chopped fresh basil
1 tablespoon red wine vinegar
4 tablespoons olive oil
salt and freshly ground black pepper

Salsas are ideal for barbecues – they're quick to prepare, excellent with grilled foods and they can be served as a sauce, dip or side salad, whatever your inclination. Tomatoes, chillies and peppers make the perfect base but you can vary the ingredients as you wish – just aim for a balance of textures, colours and flavours. Take this basic recipe as a starter and move on – this is a fiery version to go with plain grilled vegetables or to stir into salads. For those with milder tastes, the Avocado, Mango and Lime Salsa on page 82 should fit the bill.

1 Quarter, seed and chop the tomatoes. Seed and finely chop the chillies and pepper. Finely chop the onion.
2 Place all the prepared vegetables in a bowl and add the basil, wine vinegar and olive oil. Season to taste and mix well. Cover and chill until ready to serve.

Variations

AVOCADO, TOMATO AND PEPPER SALSA Add the diced flesh of 1 ripe avocado with the juice of 1 lime instead of the vinegar.

ROASTED PEPPER AND HOT TOMATO SALSA Cook the yellow pepper with 1 red pepper on the barbecue until the skin is blackened. Place in a plastic bag for 5 minutes, until the skins loosen, then peel and seed the peppers. Chop the flesh and add to the salsa. Use balsamic vinegar instead of red wine vinegar, and use ½ green chilli instead of the whole chilli. Use the best extra virgin olive oil in this version, which makes an excellent topping for bruschetta.

CUCUMBER AND CORIANDER SALSA Replace the basil with chopped fresh coriander and omit the yellow pepper. Dry fry 1 teaspoon cumin seeds and add to the salsa with 90 g (3 oz) diced cucumber.

PEANUT SATAY SAUCE

Time to prepare: 5 minutes
Time to cook: 8 minutes

Serves 4

1 small onion
1 tablespoon olive oil
1 dried chilli
2 tablespoons dark soy sauce
juice of 1 lime
1 tablespoon light muscovado sugar
8 tablespoons crunchy peanut butter
155 ml (5 fl oz) coconut milk
salt

When it comes to opinions on peanut butter, people seem to be totally divided: either they adore it or abhor it. I am of the former persuasion and this peanut butter sauce is one of my favourite recipes for cooking the humble groundnut (an alternative name reflecting the fact that peanuts grow on the ground). The sauce is based on a recipe I was given years ago in the South Pacific and it goes well with any Asian-inspired kebabs such as Vegetable Satay, see page 30. In fact, I could happily sit with a plate of cucumber strips and cubes of bread and plough my way through a whole bowlful!

1 Finely chop the onion. Heat the oil in a small pan, add the onion and cook over medium heat for about 5 minutes, until softened. Turn up the heat and continue cooking for about 2 minutes, until golden.
2 Crush the chilli and add it to the pan. Cook for a further 30 seconds. Add the soy sauce, lime juice and sugar. Remove from the heat.
3 Stir in the peanut butter, then add enough coconut milk to give the sauce a thick pouring consistency. Season to taste with salt. Serve as a sauce or dip.

Cook's Tip
You can store this sauce in a screw-topped jar in the fridge for up to a month. Do not add the coconut milk until just before serving. To use, simply heat gently and add enough coconut milk to give the correct consistency.

AVOCADO, MANGO AND LIME SALSA

Time to prepare: 10 minutes

Serves 4

1 large ripe avocado
1 large ripe mango
6 spring onions
juice of 1 lime
2 tablespoons chopped fresh mint
½ teaspoon mild curry paste
salt and freshly ground black pepper

This is a mild salsa that has a hint of spice introduced by the addition of curry paste. Serve the salsa with naan bread and the Sweet Potatoes with Quick Curry Paste, see on page 56.

1 Halve, stone and skin the avocado. Dice the flesh and place it in a mixing bowl. Peel the mango, cut the flesh away from the stone and dice it. Thinly slice the spring onions and add to the avocado with the mango.
2 Mix the lime juice, mint, curry paste and seasoning. Pour this dressing over the fruit and mix well. Serve immediately.

CUCUMBER AND WALNUT RAITA

Time to prepare: 10 minutes

Serves 4

¼ small cucumber
30 g (1 oz) walnut pieces
315 g (10 oz) Greek-style yogurt
1 teaspoon cumin seeds
few drops Tabasco sauce
2 tablespoons lemon juice
3 tablespoons chopped fresh coriander or mint
salt and freshly ground black pepper

A yogurt-based raita is the perfect accompaniment for more fiery dishes or root vegetables, which can be a little dry. I serve this recipe with practically every savoury dish I cook on the barbecue. Leave out the cumin, Tabasco and salt, add a little sugar and it even goes well with barbecued fruit!

1 Finely dice the cucumber and chop the walnut halves. Place the yogurt in a mixing bowl and stir in the cucumber and walnuts.
2 Dry fry the cumin seeds for 30 seconds, then grind to a powder in a mortar using a pestle and add to the yogurt.
3 Stir the Tabasco, lemon juice, coriander or mint and seasoning into the yogurt. Cover and chill before serving.

TOMATO CHILLI RELISH

Time to prepare: 5 minutes
Time to cook: about 25 minutes

Serves 4

750 g (1½ lb) ripe plum tomatoes or 850 g (1¾ lb) can Italian plum tomatoes
2 mild red chillies (use hot if you prefer a fiery flavour)
3 garlic cloves
60 g (2 oz) demerara sugar
4 tablespoons chopped fresh basil
salt

I first came across the original version of this sauce on a wonderful cookery day with Italian chef Carla Tomasi. It is almost like a jam, or the best-ever tomato ketchup; brush it over vegetables or tofu as they cook on the barbecue or serve a little of the sauce with kebabs or char-grilled root vegetables. Use fresh tomatoes only if they are really full of flavour, otherwise use canned Italian plum tomatoes for the correct intensity. Canned tomatoes also have the advantage of being quick to prepare whereas fresh produce has to be peeled before use.

1 If using fresh tomatoes, peel and chop them. Process the tomatoes with the chillies and garlic in a blender or food processor until smooth. Sieve the mixture into a large saucepan, pressing through as much as you can.
2 Stir in the sugar and cook over a gentle heat until it dissolves, then bring the relish to the boil and simmer for 20 to 25 minutes, until thickened. Stir in the basil and salt to taste. Serve hot or cold.

Cook's Tip
This relish can be stored in the fridge for up to a month. Pour it into warmed, sterilized pots as soon as it is removed from the heat and cover immediately with airtight lids.

SALSA VERDE

Time to prepare: 5 minutes

Serves 4

1–2 garlic cloves
1 tablespoon capers
1 small red onion
6 tablespoons chopped fresh flat-leaf parsley
4 tablespoons extra virgin olive oil
salt and freshly ground black pepper

Parsley and garlic are the essential ingredients for this, the simplest and probably the most useful, recipe in the book. I recommend flat-leaf rather than curly or English parsley for its strong flavour and the very best extra virgin olive oil. Use the sauce as a marinade or to brush over practically any vegetables as they cook on the barbecue. Spread it over toasted bread for the simplest and best bruschetta. Non-vegetarians can try this salsa spooned over mussels or any grilled fish and meat.

1 Finely chop the garlic, capers and onion. Place in a small bowl and add the parsley, oil and plenty of seasoning.

Desserts

I tend to run out of time and hot coals when it comes to adding the finishing touch to a successful barbecue. Since working with these recipes, I will aim to top up with extra coals as the barbecue progresses, in order to cook a selection of fruit which is so well suited to this cooking method.

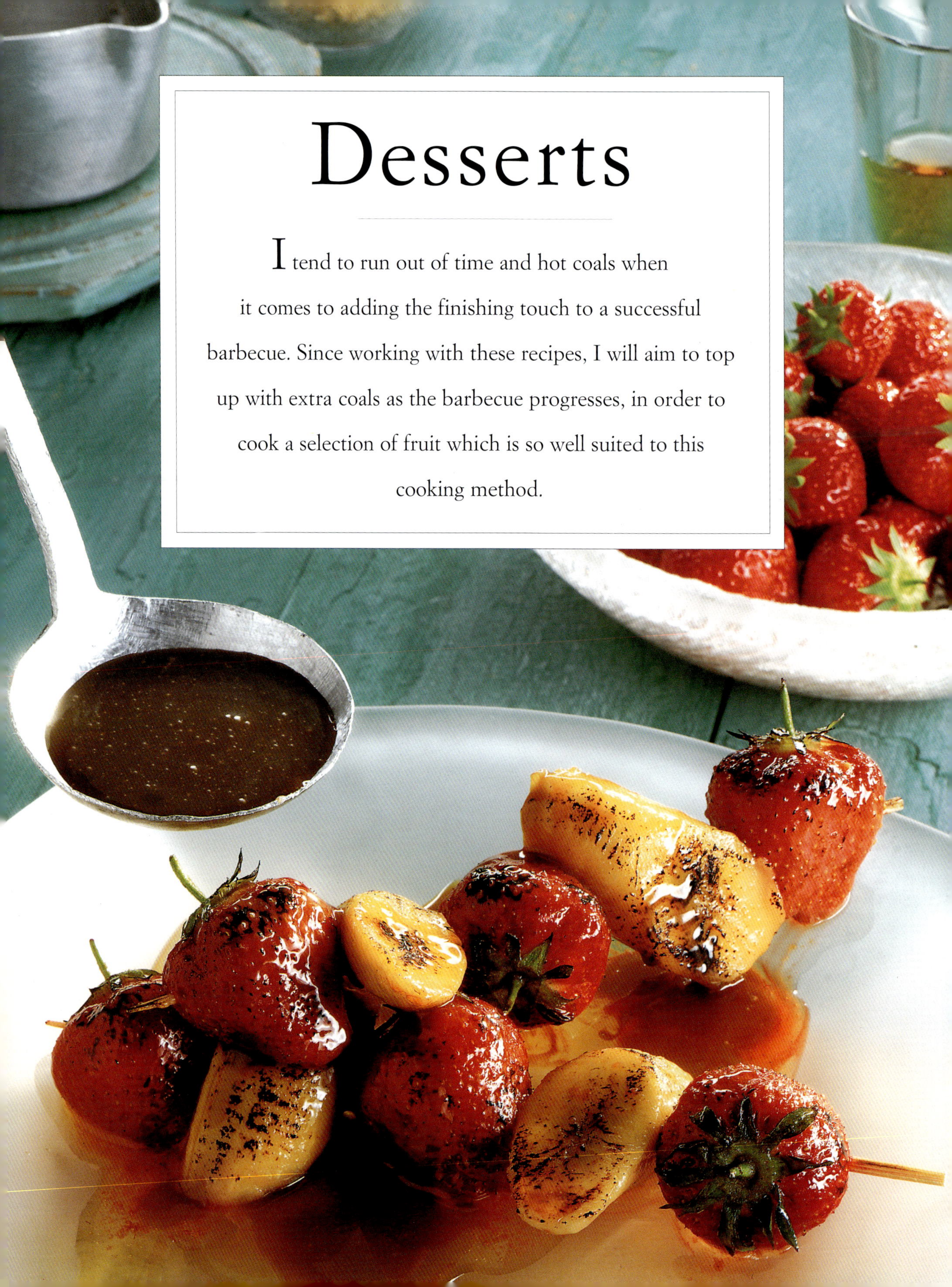

BANANA AND STRAWBERRY KEBABS WITH STICKY TOFFEE SAUCE

Time to prepare: 15 minutes
Time to cook: 5 to 8 minutes

Serves 4

30 g (1 oz) butter
1 tablespoon honey
$^1/_4$ teaspoon ground cinnamon
3 large bananas
250 g (8 oz) strawberries

Sticky Toffee Sauce
125 g (4 oz) light muscovado sugar
4 tablespoons double cream
60 g (2 oz) butter
few drops natural vanilla essence

Bananas could have been designed for barbecues. One of my stand-by finales to round off a meal cooked over the coals involves cooking bananas in their skins on the grill, turning them until they are blackened. Then the skin is cut open and spoonfuls of honey, rum and cream are poured in so that the lucky recipient can scoop out a combination of all the flavours in one go – this is incredibly simple and completely sublime! These kebabs are only slightly more time consuming to prepare and they are just as delicious.

1 Melt the butter with the honey and cinnamon. Skin the bananas and cut them into thick slices at a slant. Thread the slices onto skewers with the strawberries and brush with the melted butter mixture.
2 Make the sauce before cooking the kebabs: heat the sugar, cream, butter and vanilla essence together in a small saucepan until melted. Simmer for 5 minutes, until thickened, then set the sauce aside.
3 Cook the kebabs on the hot barbecue for 5 to 8 minutes, turning occasionally and brushing with the butter, until the bananas turn golden.
4 Serve the kebabs immediately they are cooked, with the warm sticky toffee sauce poured over.

Ladling Sticky Toffee Sauce over Banana and Strawberry Kebabs

EXOTIC FRUIT KEBABS WITH CARDAMOM RUM BUTTER

Time to prepare: 15 minutes
Time to cook: 5 to 10 minutes

Serves 4

1/2 small pineapple
1 ripe mango
1/2 paw paw

Cardamom Rum Butter
60 g (2 oz) butter
4 cardamom pods
1/2 teaspoon freshly ground black pepper
2 tablespoons light muscovado sugar
2 tablespoons dark rum or amontillado sherry

Black pepper really brings out the flavours of soft fruit; many people sprinkle it on strawberries for just that reason. Here pepper is combined with fragrant cardamom in a wonderful glaze on exotic fruits.

1 Peel the pineapple and cut it into thick slices. Remove the hard core and cut the fruit into bite-sized chunks. Peel the mango, cut the flesh away from the stone and cut it into chunks. Halve and seed the paw paw, peel it and cut it into chunks. Thread the fruit onto skewers.
2 Place the butter in a small saucepan. Crush the cardamom pods and scrape out the tiny black the seeds into the pan. Then add the pepper, sugar and rum or sherry. Heat gently until melted, then brush over the fruit.
3 Cook the kebabs on the hot barbecue for 5 to 8 minutes, turning occasionally, until the fruit is golden. Serve immediately.

BARBECUED PEACHES WITH WHISKY MAC CREAM

Time to prepare: 15 minutes
Time to marinate: 1 hour
Time to cook: 5 minutes

Serves 4

4 large ripe peaches
1 tablespoon grated fresh root ginger
½ teaspoon ground cinnamon
155 ml (5 fl oz) ginger wine

Whisky Cream
90 g (3 oz) mascarpone
155 ml (5 fl oz) double cream
2 tablespoons whisky
1 tablespoon caster sugar
grated rind of ½ lemon

I have had a liking for ginger wine ever since I was introduced to it by my Uncle Drummond, an expert in cures for all kinds of knotty problems. He ordered ginger wine as a partner for whisky in the best cure for I know of for a cold – the whisky mac. We sat together one evening in a bar in Kent and I consumed at least four of these cocktails, which certainly provided the comfort, although the resulting hangover was almost as bad as the cold it replaced! So, there tends to be a bottle of ginger wine lurking in my drinks cupboard in case of emergencies! It also makes an excellent marinade for fruit and any leftover marinade is an acceptable 'cook's perk'.

1 Pour boiling water over the peaches and leave to stand for 1 minute. Drain and skin the peaches, then halve them and remove their stones.
2 Place the peach halves in a shallow dish and scatter with the grated root ginger and cinnamon. Pour over the ginger wine, cover and leave to macerate in the fridge for 1 hour.
3 To make the whisky cream, place the mascarpone in a bowl and beat until softened. Lightly whip the cream and fold it into the mascarpone with the whisky, sugar and lemon rind. Stir in 2 tablespoons of the ginger wine from the peaches.
4 Drain the peaches and place on the barbecue. Cook the peaches for about 5 minutes, turning, and serve with the whisky mac cream.

Cook's Tip
Other fruit can be prepared using this marinade: pineapple and mango are both ideal.

PEACH AND BLUEBERRY PARCELS WITH PECAN BUTTER

Time to prepare: 10 minutes
Time to cook: 10 minutes

Serves 4

4 ripe peaches
250 g (8 oz) blueberries
30 g (1 oz) pecan halves
60 g (2 oz) butter
3 tablespoons light muscovado sugar
1 teaspoon cinnamon
4 tablespoons dark rum
fromage frais or clotted cream, to serve

I love the combination of peaches and blueberries and here they are cooked in a sweet buttery sauce that is quite divine. As you can imagine, all kinds of soft fruit can be substituted for those suggested in this recipe.

1 Halve and stone the peaches. Mix with the blueberries. Place the pecans in a small saucepan over medium heat and cook for 1 to 2 minutes, until toasted. Finely chop the pecans. Beat the butter with the sugar and cinnamon until soft and light, then beat in the pecans.
2 Cut four 25 cm (10 inch) square sheets of foil and divide the fruit between them. Dot with the cinnamon butter and sprinkle each with 1 tablespoon rum. Fold the edges of the foil and twist them to seal in the fruit.
3 Place over the coals for 10 minutes, until the fruit starts to soften. Open the foil parcels and serve with fromage frais or clotted cream.

Cook's Tip
These foil-wrapped fruit parcels can also be cooked in the oven, making them an excellent prepare-ahead dessert for dinner parties.

FIGS WITH LEMON GRASS AND HONEY

Time to prepare: 5 minutes
Time to cook: 5 to 8 minutes

Serves 4

8 ripe figs
4 tablespoons good-quality honey
1 stem lemon grass
30 g (1 oz) walnut pieces

I love figs. For me, they are forever associated with family holidays in Spain. I first sampled them picked from trees growing in the orchard around the pool of our holiday villa. Barbecuing was an everyday method of cooking the wonderful fresh ingredients from the local markets and towards the end of a meal we placed figs on the rack to cook gently in the last heat of the dying embers. Drizzled with local honey and eaten with local fresh cheese, they were the perfect way to end a summer meal.

1 Place the figs on the hot barbecue and cook for 5 to 8 minutes, until the skins start to blacken. Use tongs to turn the figs carefully as they cook, without piercing their skins. Arrange on serving plates.
2 Gently warm the honey and lemon grass in a small saucepan. Remove the lemon grass and pour the honey over the figs. Sprinkle with walnuts and serve.

LEMON-GLAZED PEARS ON RAISIN BREAD

Time to prepare: 15 minutes
Time to cook: 10 minutes

Serves 4

30 g (1 oz) butter
grated rind and juice of ½ lemon
2 tablespoons caster sugar
4 large ripe pears
4 slices cinnamon raisin bread or brioche
4 tablespoons mascarpone

My sister tested the recipes for this book and this was one of her favourites. She recommends buying the raisin bread rather than opting for brioche – in fact, she said that it was so delicious toasted on the barbecue and spread with mascarpone that she almost forgot to add the pears! Look out for cinnamon and fruit loaves in your local supermarket.

1 Place the butter in a small saucepan with the lemon rind and juice and the sugar. Heat gently until melted.
2 Peel, halve and core the pears. Brush the lemon glaze over the pears and cook them on the hot barbecue for 5 to 8 minutes, turning occasionally, until golden brown.
3 While the pears are cooking, toast the bread on both sides on the barbecue. Spread with the mascarpone and top with the cooked pears. Serve immediately.

RASPBERRY AND REDCURRANT FOOL

Time to prepare: 15 minutes

Serves 4 to 6

500 g (1 lb) fresh raspberries
125 g (4 oz) redcurrants
90 g (3 oz) caster sugar
315 ml (10 fl oz) double cream
155 ml (5 fl oz) Greek-style yogurt

Decoration
raspberries
strings of redcurrants
ginger shortbread biscuits, to serve

I make a lot of fruit fools during the summer, starting with rhubarb and gooseberry, and moving on to red soft fruit during June and July. Quick to make but always popular, they are an ideal dessert to follow a barbecued main course. Sometimes I stir in crushed meringues, which turns a strawberry fool into an Eton Mess. I serve ginger shortbread with the fool – home-made if I have time or I buy one of the high-quality varieties available at good delicatessens. Any leftover fool can be frozen to make the simplest ice cream.

1 Mash the raspberries and redcurrants together with a fork until fairly smooth but not puréed. Stir in the sugar.
2 Lightly whip the cream and stir in the Greek yogurt. Fold in the fruit purée, but do not overmix: the fool should have a slightly marbled effect.
3 Spoon the mixture into individual serving glasses and decorate with a few raspberries and strings of currants. Serve with ginger shortbread.

Cook's Tip
This recipe works well using any other soft summer fruit, such as strawberries or blackberries.

GINGER MAPLE SYRUP ICE CREAM

Time to prepare: 20 minutes
Time to freeze: 4 to 6 hours

Serves 6

4 egg yolks, size 3
4 tablespoons maple syrup
1–2 pieces stem ginger
315 ml (10 fl oz) double cream

This is a favourite recipe with all the family and I make it throughout the year as it goes as well with Christmas pudding and mince pies as it does with barbecued soft fruits. It is also good with a simple salad of one or more fresh exotic fruit, such as mango, kiwi, paw paw and/or lychees. A particular advantage of this recipe is that it does not have to be beaten part-way through freezing, so you can mix it up, freeze it and forget it. Of course it does take longer than 30 minutes to freeze the ice cream but it can be prepared several days ahead, becoming the quickest possible dessert on the day it is served.

1 Place the egg yolks in a large heatproof bowl and whisk until pale.
2 Meanwhile, heat the maple syrup in a small saucepan until just beginning to boil. Pour the hot syrup onto the egg yolks, whisking continuously, and continue whisking until the mixture is pale, thick and cool.
3 Finely chop the ginger and lightly whip the cream. Fold first the cream and then the ginger into the whisked ingredients. Pour into a freezerproof container and freeze for at least 4 hours, or until solid. Transfer to the fridge 30 minutes before serving to allow the ice cream to soften.

Cook's Tip
Make sure the whisked mixture is very thick and the ginger really finely chopped or the ginger will sink to the bottom of the ice cream.

INDEX